"*Trust in Generative Artificial Intelligence: Human-Robot Interaction and Ethical Considerations* offers a comprehensive and insightful exploration into one of the most critical aspects of AI integration today. By blending multidisciplinary perspectives with practical applications, this book serves as an essential resource for researchers, practitioners, and policymakers alike, seeking to navigate the complex landscape of AI trust and ethics."

Francesco Paolo Appio, *Paris School of Business, France*

"The development of generative artificial intelligence products like ChatGPT, Copilot, and Midjourney gives the impression that we are on the verge of achieving general artificial intelligence—a concept previously confined to science fiction. These powerful tools are transforming our society, instilling both great hopes and significant fears. How can we trust generative AI? The answer lies in this book."

Artur Gunia, *Jagiellonian University, Poland*

Trust in Generative Artificial Intelligence

Trust in Generative Artificial Intelligence: Human-Robot Interaction and Ethical Considerations is a comprehensive exploration of the critical role of trust in the rapidly evolving world of AI. Edited by Joanna Paliszkiewicz, Ireneusz Dąbrowski, and Leila Halawi, this book provides a thorough examination of how trust influences the integration and effectiveness of generative AI in various aspects of society. The book spans three main sections, offering insights from international experts across multiple disciplines and a blend of theoretical foundations and practical insights, making it a valuable resource for both understanding and implementing trust in AI systems. Readers will benefit from in-depth analyses of emerging trends, systematic literature reviews, and case studies on trust in diverse contexts, from social media to healthcare. The focus on human-robot interaction and ethical considerations provides a nuanced understanding of the psychological and ethical factors at play. Aimed at researchers, practitioners, and policymakers interested in AI, ethics, and human-robot interaction, this edited collection is also suitable for educators and students in fields related to AI and technology, providing a comprehensive resource for academic study and practical application.

Joanna Paliszkiewicz is a Professor and the Director at the Management Institute of the Warsaw University of Life Sciences, Poland.

Ireneusz Dąbrowski is an Associate Professor in the Department of Applied Economics and the Head of the Monetary Policy Unit at the SGH Warsaw School of Economics, Poland.

Leila Halawi is an Associate Professor and Research Coordinator at the College of Aviation in the Department of Graduate Studies at Worldwide Campus, Embry Riddle Aeronautical University, USA.

Routledge Studies in Trust Research
Series editors: Joanna Paliszkiewicz and Kuanchin Chen

Trust and Organisational Resilience
Iwona Otola, Marlena Grabowska and Zoran Krupka

Trust and Artificial Intelligence
Development and Application of AI Technology
Edited by Joanna Paliszkiewicz and Jerzy Gołuchowski

Trust in Social and Business Relations
Theory and Practice
Edited by Joanna Paliszkiewicz, Kuanchin Chen and Marta Mendel

Trust and Brand Management
The Role of Brand Heritage
Hanna Górska-Warsewicz

Trust and Supply Chains
Information Asymmetry in the Agri-Food Sector
Edited by Michał Pietrzak, Joanna Domagała and Aleksandra Chlebicka

Trust and Digital Transformation in the Public Sector
Barbara Kożuch and Katarzyna Sienkiewicz-Małyjurek

Trust, Media and the Economy
Mutual Relations
Edited by Joanna Paliszkiewicz, Jerzy Gołuchowski, Katarzyna Zdanowicz Cyganiak and Kuanchin Chen

Trust in Generative Artificial Intelligence
Human-Robot Interaction and Ethical Considerations
Edited by Joanna Paliszkiewicz, Ireneusz Dąbrowski and Leila Halawi

For more information about this series, please visit: www.routledge.com

Trust in Generative Artificial Intelligence

Human-Robot Interaction and Ethical Considerations

Edited by
Joanna Paliszkiewicz,
Ireneusz Dąbrowski and Leila Halawi

Routledge
Taylor & Francis Group
NEW YORK AND LONDON

First published 2025
by Routledge
605 Third Avenue, New York, NY 10158

and by Routledge
4 Park Square, Milton Park, Abingdon, Oxon OX14 4RN

Routledge is an imprint of the Taylor & Francis Group, an informa business

ISBN: 978-1-032-95794-4 (hbk)
ISBN: 978-1-032-95876-7 (pbk)
ISBN: 978-1-003-58693-7 (ebk)

DOI: 10.4324/9781003586937

Typeset in Times New Roman
by codeMantra

For my daughter, Natalia—may this book inspire you to dream boldly and pursue your passions with courage

JP

To my children, Weronika and Irek, my forever inspiration

ID

In memory of my father

LH

Contents

Tables

Figures

Editors

Joanna Paliszkiewicz works as a Full Professor at the Warsaw University of Life Sciences (WULS—SGGW). She is the Director of the Management Institute. She is also an Adjunct Professor at the University of Vaasa in Finland. She obtained the academic title "Full Professor" from the International School for Social and Business Studies in Slovenia. She is well recognized in Poland and abroad for her expertise in management issues: knowledge management and trust management. She has published over 250 papers/manuscripts and is the author/co-author/editor of 21 books.

She has been a part of many scholarship endeavors in the USA, Ireland, Slovakia, Taiwan, the UK, and Hungary. She has actively participated in presenting research results at various international conferences. Currently, she serves as the Deputy Editor-in-Chief of the *Management and Production Engineering Review*. She is an Associate Editor for the *Journal of Computer Information Systems, Expert System with Applications*, *Issues in Information Systems, Przegląd Organizacji*, and *Intelligent Systems with Applications*. She is the Vice President of the Polish Association for Production Engineering. She also serves as the Chair of the International Cooperation in European Business Club. She serves as the Vice President of the International Association for Computer Information Systems in the USA. She is a board member of the Intellectual Capital Accreditation Association. She has successfully supervised many PhD students leading them to completion of their degrees. She has also served as an external reviewer for several PhD students in Poland, India, and Finland. She is actively involved in participating in the scientific committees of many international conferences. She was named the 2013 Computer Educator of the Year by IACIS in the USA.

Ireneusz Dąbrowski, PhD, is an Associate Professor in the Department of Applied Economics and the Head of the Monetary Policy Unit at the SGH Warsaw School of Economics. He is a polish economist and lawyer who served as Undersecretary of State in the Ministry of the Treasury of the Republic of Poland from 2006 to 2007 and as a member of the Monetary

Policy Council of the National Bank of Poland from 2022. He is an architect of Poland's privatization reforms. He graduated from the SGH Warsaw School of Economics with a Master's degree in Finance and Banking in 1998, then from Warsaw University with a Master's degree in Corporate Law in 2002. He received PhD in Economics from the SGH Warsaw School of Economics in 2006 after completing and defending his disser tation on General Equilibrium Theory. He habilitated in Economics in 2017. He is a Visiting Scholar at the College of Business and Management – Northeastern Illinois University in Chicago, is a member of the Polish Economic Society and the Illinois Economic Association, and author of popular books and scientific articles in the field of economics.

Dr. Leila Halawi is an esteemed tenured Associate Professor and Research Coordinator at the College of Aviation, Department of Graduate Studies, Worldwide Campus. With a rich history of academic leadership, she has served as the Program Chair for the Master of Science in Aviation and Aerospace Sustainability (AASI) program and the Graduate Certificate in Aviation and Aerospace Sustainability. Additionally, she has held key roles such as the Master of Science in Management Information Systems (MMIS) Program Chair, MIS Discipline Chair, Chair of the Research Committee for the Senate, and Director of Research for the College of Business. Her contributions extend beyond administrative roles; she has been instrumental in curriculum development, notably designing the MMIS curriculum and forging valuable partnerships with industry leaders such as SAS®, Tableau, and Microsoft. These alliances have resulted in groundbreaking initiatives, including a joint Business Intelligence and Analytics specialization certificate with SAS. Recognized for her commitment to excellence in education, she holds certifications in online teaching from Sloan C Consortium and Quality Matters. She is also a certified DACUM facilitator and coordinator, leveraging her expertise to enhance curriculum design and industry relevance.

In the realm of academia, her impact is widespread. She serves as an editorial review board member for the Enterprise Arciihitecture Body of Knowledge (EABOK) at Mitre Corporation and as a reviewer for esteemed journals and conferences, including the *Journal of Computer Information Systems* (JCIS) and the European Conference on Social Media (ECSM). Her dedication to research and scholarship is evident in her numerous publications and accolades, including the SAS 2020 Emerging Educator Award and co-authorship of influential analytics and predictive analytics books. Her research interests encompass a broad spectrum of topics, from knowledge management to predictive analytics, reflecting her commitment to addressing industry needs and fostering innovation. She leads a pioneering research endeavor funded by a NIST grant to integrate engineering and aviation standards into higher education curricula.

Beyond academia, she enjoys a vibrant life in Clearwater, Florida, indulging in passions such as traveling, swimming, and participating in national and international conferences. Her multifaceted contributions continue to shape the landscape of academia and industry, bridging the gap between education and real-world demands.

Contributors

Agata Berdowska is a researcher and lecturer at the University of Economics in Katowice. She started her work in the Department of Knowledge Engineering. Since 2017, she has been a member in the Department of Communication Design and Analysis. Her subject of scientific interest is concentrated around new information technologies: technologies supporting fact-checking, conversation systems (chatbots) and electronic training (e-learning), and the labor market, new ICT professions. She has participated in two international projects: "European Cooperation Project on Disinformation and Fact-Checking Training" (2019–2022), the "CALYPSO – Collaborative Fact-Checking" (2021–2022). She is the co-author of the chapter titled "Building trust in professional business journalism in the age of digital media and disinformation spreading" in the book *Communication, Leadership and Trust in Organizations* edited by Joanna Paliszkiewicz, José L. Guerrero Cusumano and published in 2023 (Routledge). She is also the co-author of the chapter titled "Digital Therapists – the Potential and Limitations of Artificial Intelligence" in the book *Trust and Artificial Intelligence. Development and Application of AI Technology* edited by Joanna Paliszkiewicz, Jerzy Gołuchowski published in 2024 (Routledge).

Fatih Çetin is a Professor of Management in the Department of Business Administration at Baskent University in Turkey. His research focuses on the field of organizational behavior and knowledge management studies. His recent researches have appeared in the *European Journal of International Management*, *International Business Review*, and *Decision Analytics Journal*.

Anna Dąbrowska is a Full Professor in the Department of Consumer Behaviour Research, Collegium of Management and Finance, Warsaw School of Economics, Poland. Her research interests focus on the services market, e-services, consumption of goods and services in households, consumer behavior in the market of goods and services, and protection of consumer rights. She is the author of many scientific and popular science publications published in Poland and abroad.

Dr. Alicja Fandrejewska is an Adjunct/Associate Professor at the Warsaw School of Life Sciences, Institute of Management. She has worked at the Warsaw School of Life Sciences, Kozminski University, and the University of Warsaw. She has a PhD in Social Sciences (Management) awarded by the Warsaw School of Economics and a PhD in Humanities (English Philology) obtained from the University of Gdańsk. Her research focuses on the development of ICT technologies, examining their usability for consumers and their role in enhancing the efficiency and competitiveness of businesses in a globally interconnected, knowledge-based economy. She is also an author and co-author of publications in the areas of cross-country comparisons of acceptance of technological innovations, including e-health, cultural differences, business communication, ICT, and management.

Jerzy Gołuchowski is a Full Professor at the University of Economics in Katowice. He has served as Vice-Rector (2005–2009) and Dean (2009–2019, 2020–now) of the Faculty of Informatics and Communication and is the Head of the Department of Communication Design and Analysis (since 2017) His scientific interests are concentrated around communication issues in decision-making processes, artificial intelligence, and management of organizations, particularly the management of trust, knowledge, and communication in organizations. His publications include B. Filipczyk and J. Gołuchowski, Cyfrowa komunikacja organizacji [*Digital Communication of Organization*] (2020), and J. Liebowitz, J. Paliszkiewicz, and J. Gołuchowski (eds.), *Intuition, Trust, and Analytics* (Taylor & Francis, 2018). He is the co-author of the chapter titled "Building trust in professional business journalism in the age of digital media and disinformation spreading" in the book *Com munication, Leadership and Trust in Organizations* edited by Joanna Paliszkiewicz, José L. Guerrero Cusumano and published in 2023 (Routledge). He has participated in international projects such as the "CALYPSO – Collaborative Fact-Checking" (2021–2022), "European Cooperation Project on Disinformation and Fact-Checking Training" (2019–2022), "Language Communication on eCommerce Websites" (2004–2007), and "European Master in e-Commerce and Cyber Communication" (2000–2003). He is a member of the Society for Social Communication and Scientific Society for Business Informatics. He has also served as President of the Scientific Society for Business Informatics and as a member of the Council of Scientific Societies of the Polish Academy of Sciences. From 2012 to 2016, he was a member of the Programme Council of TVP o/Katowice and, from 2014 to 2016, a member of the Supervisory Board of Polish Radio Katowice.

Hanna Górska-Warsewicz is an Associate Professor at the Management Institute of the Warsaw University of Life Sciences. Habilitated at the Faculty of Management, University of Warsaw, she specializes in the subject of brands, brand equity, brand identity, brand preference, and brand management.

She is the author or co-author of over 150 publications in this field, including eight books, among others, *Trust and brand management. The role of brand heritage* (Routledge Taylor and Francis) and *Reputation and equity of corporate brand. A dynamic and multidimensional approach* (PWN Edition).

Sam Holley, PhD, an accomplished educator and researcher, has taught over 60 courses and chaired 600+ research projects and dissertations at the university level. His diverse career includes key positions at the White House, Pentagon, and academia, where he received numerous awards for teaching and research. A licensed clinical neuropsychologist and certified pilot/instructor, His expertise spans neuropsychology, human factors, aerospace medicine, and aviation.

Anna Jasiulewicz is an Assistant Professor at the Management Institute of the Warsaw University of Life Sciences, the Head of Postgraduate Public Procurement Studies, and a member of the Polish Scientific Society of Marketing. She is a researcher in international and national projects, author of several dozen scientific publications, organizer of international conferences, and reviewer of scientific articles in national and international journals. She specializes in issues related to digital marketing, consumer behavior, and sports marketing.

Ewa Jaska, PhD in Economic Sciences, is an Assistant Professor at the Institute of Economics and Finance of the Warsaw University of Life Sciences. Her main areas of scientific interest are mass communication, media market and media management, and enterprise communication with the market. She is the author and co-author of more than 140 publications, including two books *Economic and Social Determinants of Changes in the Polish Television Market* and *Communication of the Enterprise with the Environment in the Knowledge-based Economy*, scientific editor of 12 collective works, including three volumes of the series "Media in the information society"—member of expert teams and contractors in research projects. Currently, she is a member of the research team of the project "Environmentally and socially sustainable behaviour of Generation Z stimulated by mobile applications" under the "Science for Society" program.

Xuan Tai Mai is a lecturer at the FPT University in Vietnam. His main research interests are marketing, e-commerce, social media, and consumer behavior. He has published in the *International Journal of Information Management*, *International Journal of Tourism Policy*, *International Journal of Business Information Systems*, and *Journal of Systems and Information Technology*.

Josip Marić, PhD, is an Associate Professor and a member in the Department of Supply Chains and Digital Management at EM Normandie Business School, France. His research interests are in the areas of innovation management and

Management of Information Systems (MIS), focusing on the topics of digital transformation, implications of digital technologies, and Industry 4.0. His work is presented at various scholarly events in France and abroad. At the same time, major contributions appear in internationally recognized journals such as *Technological Forecasting and Social Change*, *Technovation*, *Annals of Operations Research*, *Journal of Enterprise Information Management*, *International Journal of Logistics Research and Applications*, *Journal of Manufacturing Technology Management*, and *Global Journal of Flexible Systems Management*.

Dr. Angelica Marotta, as a renowned cybersecurity professional, boasts an international research, consulting, and teaching portfolio. Her interdisciplinary background is rooted in a Bachelor's degree in Computer Science from the University of Pisa, a Graduate Certificate in Cyber Security, a Master's degree in Justice Studies with a specialization in cybersecurity from Southern New Hampshire University, and a PhD in Cybersecurity from the University of South Wales, UK. She also holds the ISO/IEC 27001:2013 certification as a Certified Information Security Management System Auditor. Currently working as a researcher in the Cybersecurity at MIT Sloan (CAMS) Research Consortium at the Massachusetts Institute of Technology (MIT), her research primarily focuses on cybersecurity regulatory compliance. Her significant contribution to the field is also reflected in her additional collaborations with top-tier organizations, such as the World Economic Forum (WEF) and the European Cyber Security Organisation (ECSO), where she is engaged in high-impact projects. As an esteemed author, she has published over 30 scholarly articles in prestigious cybersecurity journals and contributed chapters to leading industry books. Her academic works cover a wide range of topics, including cyber insurance, risk management, cybersecurity culture, robotics, big data analytics, artificial intelligence, and compliance. Her expertise extends beyond research, as she is also a seasoned educator. She has taught and developed courses, along with providing dissertation supervision and training at universities in the USA and UK. Her academic endeavors in the field have earned her outstanding recognition, including the 2021 Best Paper Award at the IACIS Europe Conference. She is also a frequently sought-after keynote speaker and guest lecturer at global cybersecurity conferences and events.

Marta Mendel works as an Adjunct Professor at the Warsaw University of Life Sciences (WULS—SGGW). She is the vice-rector of international cooperation and former dean of the Faculty of Veterinary Medicine at the Warsaw University of Life Sciences. She is well recognized in Poland and abroad for her expertise in veterinary pharmacology and toxicology. Her main research focus includes the identification of pharmacological activities of medicinal plants and phytoconstituents, as well as the development of new alternative models for pharmacological studies. She authored and co-authored over 60 original papers with a total IF over 70. She is an expert of the Polish Accreditation Agency, Polish National Agency of the Academic Exchange, and the International Evaluation Programme implemented by the European University Association. She co-chairs the networking group on Medicinal Plants and Natural Products in Animal Healthcare and Veterinary Medicine of the Society of Medicinal Plants and Natural Product Research. Currently (2023–2024), she serves as the Co-Chair of the Board of Directors of the Green European University. She is supervising four PhD students. She has also served as an external reviewer for several PhD and post-doctoral theses. She is actively involved in participating in the scientific committees of national and international conferences. In addition to didactic and research projects, she is engaged in developmental aid.

Dr. Mark Miller, EdD, leads the Master of Science in Aeronautics Program at Embry-Riddle Aeronautical University Worldwide Campus, specializing in aviation safety, human factors, and education. His research focuses on flight deck interfaces, SMS human factors in aviation maintenance, and organizational influences on safety. Dr. Miller contributes to curriculum development, consults with airlines on safety and human factors, and partners in a NIST-funded research project on integrating engineering and aviation standards into higher education

Etleva Muça, Professor, is a scientific researcher and lecturer of Sustainable Rural Development, Spatial Planning, and EU Territory and Cohesion Policies at the Agricultural University of Tirana, Albania—Faculty of Economy and Agribusiness since September 2001. Her main research interests focus on sustainable local and regional development (territory development, social capital, rural tourism, and institutions and consumer choices).

Joanna M. Moczydłowska is a Professor in the field of social sciences, an employee of the Faculty of Management Engineering at the Bialystok University of Technology, the Head of the Chairs of Management, Economics and Finance, and specializes in human capital management issues, including the use of new technologies in HRM, as well as organizational behavior and business psychology.

Anna Musialik-Chmiel is an Assistant Professor in the Department of Communication Design and Analysis at the University of Economics in Katowice, specializing in Journalism and Media Studies. The main fields of academic interest are mission of public broadcasting, critical analysis of media coverage, business and economic journalism, and fact-checking. She cooperates with the public media (Polish Radio Katowice). She is the co-author of the chapter titled "Building trust in professional business journalism in the age of digital media and disinformation spreading" in the book *Communication,*

Leadership and Trust in Organizations edited by Joanna Paliszkiewicz, José L. Guerrero Cusumano and published in 2023 (Routledge).

Trang Nguyen is currently working as a lecturer in the School of International Business and Marketing, University of Economics Ho Chi Minh City, Vietnam. She received her Doctoral degree from the Institute of International Management, College of Management, National Cheng Kung University, Taiwan. Her studies have appeared in Information and Management, Computers in Human Behavior, Internet Research, *Journal of Electronic Commerce Research, Journal of Internet Commerce, Online Information Review, International Journal of Business Information Systems, Journal of Systems and Information Technology*, and several international conference proceedings.

Irena Ozimek is a Full Professor at the Management Institute of the Warsaw University of Life Sciences, Poland, and works as a scientific researcher and lecturer. Her research focuses on consumer protection (including consumer education), consumption of goods and services in households, and consumer behavior in the market of goods and services, especially in the food market and tourism. She is the author of many scientific and popular science publications published in Poland and abroad.

Vesna Paraušić, PhD, is employed as a senior research associate at the Institute for Agricultural Economics in Belgrade, Serbia. She mainly deals with economic, environmental, and social issues in the sphere of agriculture and rural development. Also, her research interests focus on strategic planning of sustainable development of local communities and regions, social capital, quality of life, as well as local development initiatives. She is a member of the Scientific Association of Agrarian Economists of the Balkans (NDAEB), the Association of Agrarian Economists of Serbia (DAES), the European Rural Development Network (ERDN/European Rural Development Network), as well as the scientific research network RebResNet.

Mirjana Pejić-Bach, PhD, is a Professor in the Department of Informatics, Faculty of Economics in Zagreb (Croatia). She holds a PhD in System Dynamics Modeling from the Faculty of Economics, University of Zagreb, and was trained at the MIT Sloan School of Management in System Dynamics. Her research areas are the strategic application of information technology in business, data science, and simulation modeling. She has won several international awards for her scientific work, such as the Emerald Literati Network Awards for Excellence, and has published in recognized journals such as *International Journal of Information Management, Baltic Journal of Management, Systems Research and Behavioral Science, Kybernetes, Business Process Management Journal*, and *International Journal of Engineering Business Management.*

Iwona Pomianek, PhD, is a scientific researcher and lecturer at the Institute of Economics and Finance, Warsaw University of Life Sciences – SGGW, Poland. Her main research interest is focused on sustainable local and regional development, including development potential, local government activity, strategies, entrepreneurship, infrastructure, quality of life, tourism attractiveness, use of the EU funds.

Wojciech Pizło is a Professor at the Management Institute of the Warsaw University of Life Sciences (2021–); previously he was also employed as a Professor at the Institute of Management, Jan Kochanowski University in Kielce (2012–2018) and the Institute of Economics and Finance (2014–2021) of the Warsaw University of Life Sciences. His main interests are in management in the era of the information revolution, in particular marketing theory, institutional economics, leadership, and creativity. He is the author of monographs such as *Enterprise in the Information Society in the Light of the Theory of Institutional Economics* (2009), *Marketing. Economic Knowledge and Market Activity* (2017), *International Marketing – the European Point of View* (2019), as well as the author and co-author of a thesis on virtual organizations in international research teams, the use of AI in marketing activities and in insurance companies. He is the author of over 150 articles and chapters in monographs, scientific, and thematic editor in scientific journals.

Liudmyla Shulhina is a Professor of Economics at the National Technical University of Ukraine "Igor Sikorski Kyiv Polytechnic Institute". He is the author of many scientific and popular science publications published in Ukraine and abroad. Currently, his research interests focus on the issues of business communication, performance marketing, digital strategies of the company, and consumer behavior.

Edyta Skarzyńska is research and teaching assistant at the Management Institute of the Warsaw University of Life Sciences. Her scientific interests focus on topics related to intellectual capital, social media, as well as the use of artificial intelligence and trust in the field.

Selay Ilgaz Sümer is an Associate Professor in the Department of Business Administration, Faculty of Economics and Administrative Sciences, Başkent University, Ankara, Turkey. She has published several articles in different journals and been indexed in sources like Social Sciences Citation Index (SSCI), Science Citation Index-Expanded (SCI-Exp), Crossref, and Index Copernicus. She has published numerous articles, conference proceedings, and book chapters.

Agnieszka Tul-Krzyszczuk, PhD, is an Assistant Professor at the Management Institute of the Warsaw University of Life Sciences, Poland. She is a consultant, trainer, and educator specializing in the organization and management of production and service enterprises, particularly in the food, hotel, and food-service industries. She conducts scientific research in the fields of management, marketing, innovation, and digital trust. She actively participates in national and international grants and projects. Furthermore, she is the author of numerous scientific and popular science publications in both national and international journals. She has also presented her work at various national and international conferences.

Agnieszka Werenowska, PhD in Economic Sciences, is an Assistant Professor at the Management Institute of the Warsaw University of Life Sciences. The mainstream research includes public relations, media communication, marketing in enterprises and local government units, and promotion of sustainable consumption. She is the author and co-author of 120 scientific publications published in Polish and foreign scientific journals, co-author of the monograph *Communication of an Enterprise with the Environment in a Knowledge-Based Economy*, co-editor of the textbook *Recent Trends, Concepts and Methods in Management*, member of the research team of the project "Environmentally and socially sustainable behaviour of Generation Z stimulated by mobile applications" under the "Science for Society" program.

Urszula Widelska, PhD, is a Professor at the Faculty of Social Sciences and Humanities of the University of Lomza; a Deputy Head of the Department of Economics and Finance; author of dozens of publications, mainly in the field of relationship marketing and customer orientation; leader and expert in numerous research projects, including grants from the National Science Centre; an Editor-in-Chief of the scientific journal *Marketing and Market* published by the PWE.

Forewords

AI has evolved through various waves in computing history. Generative AI, a significant advancement emerging from the field of Natural Language Processing, has set a new milestone in AI research. Following many consumer-facing AI technologies (e.g., Siri, photo tagging, and product recommendations), today's generative AI applications are designed for widespread accessibility, making them useful even to non-technical audiences. Every day, new use cases are discovered and shared across the Internet, highlighting both opportunities and risks associated with this technology. Consequently, scholarly discussions about generative AI should extend beyond only its potential benefits to include its broader societal and ethical implications. Unlike those AI applications that were embedded in major technology ecosystems (e.g., Apple's Siri), generative AI quickly attained widespread ubiquity unmatched by previous standalone technologies before it was incorporated into other technology systems. This widespread reach means that concerns relevant to other consumer technologies are increasingly applicable to generative AI as well.

This book begins by laying a strong foundation to help readers engage with more complex topics related to human-AI interactions and contextual applications. The opening section provides an overview of key research and explores societal impacts through real-world usage scenarios. A crucial factor in the success of consumer-facing AI is its ability to generate meaningful insights through user interactions, a theme explored throughout the text. The second section delves into human-AI interaction, examining trust, ethics, and psychological drivers of AI usage. Finally, this book concludes with a focus on cross-cultural comparisons, generational differences, and specialized applications of generative AI.

In summary, this book addresses both the positive and negative consequences of generative AI, covering outcome-based assessments such as risks, brand equity, and ethical concerns, as well as process-oriented factors like communication dynamics, psychological influences, and confidence building. This comprehensive approach is essential as the relevance of generative AI extends far beyond purely technical discussions. The research presented is both timely and applicable to a broad audience, offering readers a journey from fundamental concepts to practical implementations. Carefully curated articles and a well-structured narrative ensure that the content captivates both academic and professional audiences.

Kuanchin Chen, John W. Snyder Fellow
Director of Center for Business Analytics
Professor of Computer Information Systems and Business Analytics
Western Michigan University, USA

Forewords

Since this book focuses on generative AI (GenAI), I would like to thank ChatGPT for helping me write this Foreword. In recent years, artificial intelligence has progressed beyond mere automation and data analysis to encompass generative capabilities that can produce content with remarkable depth and complexity. This shift heralds a new era in human-technology interaction, where the lines between human and machine-generated output are increasingly blurred. As generative AI systems become more sophisticated, they offer the potential to revolutionize industries, enhance creativity, and address societal challenges. However, with these advancements come unprecedented ethical concerns and a pressing need to establish trust.

Trust in Generative Artificial Intelligence: Human-Robot Interaction and Ethical Considerations could not be timelier. This volume, edited by Joanna Paliszkiewicz, Ireneusz Dąbrowski, and Leila Halawi, presents a vital exploration of the many dimensions of trust that underlie our interactions with AI. This work not only examines theoretical constructs of trust and ethical considerations but also delves into specific applications across domains such as healthcare, aviation, and digital media, highlighting the varied roles trust plays in our acceptance of AI technologies.

As we view ethics and trust in the GenAI world, we must also pay closer attention to regulations, policies, or guardrails that are needed, especially as related to hate speech created by GenAI (Liebowitz, 2025). GenAI systems can hallucinate and many people (and students) take and "trust" what they say at face value. We must further educate society about these potential pitfalls, and this book provides some help in this regard.

This book is structured in three parts, each providing unique insights into the formation, challenges, and applications of trust in generative AI. From the foundations of trust to ethical and psychological factors in human-robot interactions, and finally, to practical applications across different sectors, this volume provides a holistic approach to understanding trust in a rapidly evolving technological landscape. It offers readers the tools to critically assess generative

AI's role in society and equips them to make informed decisions regarding its integration.

As you engage with this book, you will encounter perspectives from experts across disciplines and geographies, each bringing their unique lens to the ethical and practical challenges that generative AI poses. These insights make this volume not only a foundational text for researchers, practitioners, and students but also a valuable resource for anyone seeking to navigate the complexities of trust in AI.

Trust is a cornerstone of successful human-AI collaboration, and as we step further into a world shaped by generative technologies, it is essential that we consider how to foster and maintain it. I am confident that this book will inspire a thoughtful approach to these considerations, helping readers appreciate the power of trust as we chart the future of AI.

Reference

Liebowitz, J. (ed.) (2025), *Regulating Hate Speech Created by Generative AI*, Taylor & Francis.

Jay Liebowitz, D.Sc.
Professor of Business Innovation and Industry Transformation
Director, AI-EDGE Center
Crummer Graduate School of Business
Rollins College
Winter Park, Florida, USA

Preface

As artificial intelligence (AI) reshapes our daily lives, its rapid growth raises crucial questions about trust and the role of ethics in human-technology interactions. Among AI's most transformative advancements, generative AI—systems capable of producing complex, human-like content—stands at the frontier, redefining the ways we perceive and engage with technology. This accelerated integration of generative AI into diverse sectors presents both opportunities and challenges, especially as society grapples with understanding the foundations of trust and accountability in our interactions with these systems. The impact of generative AI on human-robot interaction, its ethical considerations, and the implications for society form the basis of this comprehensive exploration.

Trust in Generative Artificial Intelligence: Human-Robot Interaction and Ethical Considerations explores these complex questions. Edited by Joanna Paliszkiewicz, Ireneusz Dąbrowski, and Leila Halawi, this book brings together leading voices from multiple disciplines to address the psychological, ethical, and practical dimensions of trust in generative AI. Our contributors delve into the theoretical foundations of trust, examining both the potential risks and ethical dilemmas posed by AI technologies. They offer insights into human-robot interaction and the various determinants that influence our ability to trust these systems.

Divided into three parts, this book provides a structured journey through the foundations, ethical considerations, and context-specific applications of trust in generative AI. Readers are introduced to concepts of trust within human-robot interaction, the ethical frameworks necessary for AI deployment, and the unique challenges trust presents in fields as varied as healthcare, social media, and commercial aviation. Each chapter represents a comprehensive study by experts from around the world, collectively building a nuanced understanding of the role trust plays in successful human-AI collaboration.

This book aims to serve researchers, practitioners, and policymakers who are striving to navigate the ethical landscape of AI, as well as educators and students seeking a robust foundation on the topic. In a world where AI continues to shape the future, we believe that cultivating trust in these technologies is not merely an academic endeavor but a societal imperative. It is our hope that this volume will inspire a deeper understanding of the ethical considerations and foster informed trust, enabling generative AI to act as a positive force in our lives.

Joanna Paliszkiewicz, Warsaw University of Life Sciences, Poland,
University of Economics Ho Chi Minh City (UEH), Vietnam
Ireneusz Dąbrowski, Warsaw School of Economics, Poland
Leila Halawi, College of Aviation, Embry Riddle Aeronautical University, Florida, USA

Part I

Foundations of trust in generative AI

1 Risks, failures, and ethical dilemmas of AI technologies and trust

Edyta Skarzyńska, Joanna Paliszkiewicz, Ireneusz Dąbrowski and Marta Mendel

Introduction

Artificial intelligence (AI) has been undergoing rapid development in recent years, significantly impacting daily life, the economy, and the broader functioning of modern society. AI-based technologies can potentially transform many sectors, from health care to education to industry and commerce. The application of AI in areas such as big data analytics, image recognition, natural language processing, and process automation can bring numerous benefits and improvements, impacting productivity, time, and resource savings (Afroogh et al., 2024). However, the rapid development and growing use of AI also brings risks, challenges, and ethical dilemmas that can have serious consequences at both the individual and system levels. The purpose of the chapter is to identify key risks and failures associated with AI technologies as well as to discuss significant ethical dilemmas that may affect public trust in these technologies. In particular, attention is paid to problematic aspects of the operation and use of AI that can threaten security, privacy, and social justice (Dirgová Luptáková et al., 2023). The development of AI technology is not only a technological issue but also one deeply rooted in social and ethical contexts, making it necessary to consider these aspects in the design, implementation, and regulation of such systems (Kochetova, 2024). The importance of this study stems from the growing role that AI plays in modern societies and economies. With the increasing trust that people place in AI technologies, there are also many questions about their reliability, ethics, and the implications of their use. While the technology promises improvements in efficiency and innovation, it also raises questions about the transparency of decisions over which users and those affected by AI often have no control (Al-kfairy et al., 2024).

This chapter also reviews the literature on the risks, mistakes, and ethical dilemmas associated with AI technologies. Many researchers and experts point to the need to develop AI sustainably and to ensure that systems based on this technology are designed and implemented by ethical principles (Bostrom & Yudkowsky, 2018). Understanding and addressing these issues is crucial, as AI technologies will continue to shape the future of societies and economies.

DOI: 10.4324/9781003586937-2

In the face of dynamic technological change and the increasing impact of AI on everyday life, a consistent and thoughtful ethical and regulatory framework is needed to minimize the risks and negative impacts associated with this technology (Guan et al., 2022).

Characteristics of risks and failures of artificial intelligence technology

The deployment of AI brings significant benefits to many sectors of the economy, including improving operational efficiency, optimizing costs, and fostering innovation. However, several risks and failures can affect the effectiveness of the application of these technologies, as well as the financial stability of organizations and the security of their stakeholders (Afroogh et al., 2024). A look at the main risks, a case study of failures, and their consequences provide a better understanding of how AI challenges can affect the effective management of organizations and the economy (Bostrom & Yudkowsky, 2018).

One of the biggest risks of implementing AI is the unpredictability of its performance in certain situations, especially under conditions that differ from the training data on which the models were taught (Kochetova, 2024). AI models are designed based on analysis of historical data but are often unable to adapt to new, changing conditions or situations. From a management perspective, this means that companies relying on such systems face the risk of erroneous results, which can lead to unexpected financial and reputational consequences (Treacy, 2023).

The quality of the data used to train AI models is crucial to their correct operation. Low-quality, incomplete, or error-prone data can lead to malfunctioning models and, consequently, wrong decisions. From an economic perspective, this risk is significant, as companies may incur high costs associated with having to correct decisions or implement additional safeguards at the data processing stage. From a management perspective, this problem requires the introduction of appropriate data quality control procedures and ongoing monitoring of the results generated by the models (Polemi et al., 2024). AI algorithms are prone to errors due to the limitations of the algorithms themselves or to mistakes made at the implementation stage. Additionally, biases in training data can cause AI models to generate biased results, which can lead to unfair treatment of customers or users. In the context of management and economics, the consequences of such biases can be far-reaching. For example, algorithmic biases can result in discrimination in hiring processes or credit evaluations, which negatively affect a company's reputation and can lead to legal consequences.

AI systems often require access to large sets of personal data, which carries the risk of violating individual privacy (Afroogh et al., 2024). From an economic and management perspective, data privacy issues can result in hefty financial penalties due to regulations such as the GDPR. In addition, breaches of customer privacy can lead to a loss of customer trust, which affects customer relationships and an organization's long-term financial performance (Guan et al., 2022).

AI failures have a variety of consequences, both at the financial and social levels. In the case of unpredictable performance or algorithmic errors, organizations can suffer financial losses related to correcting decisions and having to adapt systems to new conditions. For example, companies in the financial sector may face losses due to improper credit risk assessment, which can lead to inefficient allocation of financial resources. The social impact is equally important, as AI failures can lead to a breach of public trust, which has a direct impact on an organization's long-term reputation and customers' propensity to use its services (Polemi et al., 2024). Additionally, algorithmic biases in recruitment or credit evaluation can exacerbate social inequalities, which affects individuals and can lead to increased social costs associated with the need to level the playing field. In the case of errors in health systems, these consequences can be particularly severe, even leading to risks to patients' health and lives. In the face of such consequences, from a management perspective, it is necessary to put in place appropriate procedures for monitoring and quality control of IS systems to catch irregularities on an ongoing basis and prevent potential risks to individuals and communities. Effective IS risk management requires not only knowledge of potential risks but also the implementation of appropriate procedures that minimize the risk of failures and support the building of trust in IS (Afroogh et al., 2024).

Ethical dilemmas related to AI and trust

AI technologies offer tremendous potential, but their development and implementation bring with them numerous ethical dilemmas that can affect the level of public trust in these technologies (Russell & Norvig, 2016). The introduction of AI into the sphere of social, economic, and political life raises questions about accountability, transparency, and respect for individual values such as dignity and privacy. With the growing number of AI applications, several ethical issues arise due to the characteristics of the algorithms themselves, how they are created, and the data on which they are trained. One of the main problems is the lack of transparency in the algorithms. AI-based algorithms, especially those using deep learning, often function as so-called black boxes – their internal mechanisms are difficult to understand even for specialists, making their decisions potentially unclear to users. The inability to explain why an algorithm made a certain decision raises doubts about its reliability and fairness. This is especially problematic when algorithm decisions involve key aspects of people's lives, such as the allocation of credit, the evaluation of job candidates, or access to health services (Treacy, 2023).

Another major ethical dilemma is the difficulty of assigning responsibility for AI decisions. In classical decision-making systems, responsibility is easier to assign to specific individuals or organizations. In the case of AI, however, where decision-making is distributed and based on data and algorithms created by many different developers, the question of responsibility becomes more complex (Polemi et al., 2024). Who should bear the consequences for erroneous AI decisions – the algorithm creator, the data provider, the company implementing the AI, or perhaps the tool itself? The lack of a clear answer to this question makes it difficult to create a legal framework and appropriate ethical rules, which in turn can lead to a situation in which aggrieved individuals are left without any means of seeking justice.

Another key ethical issue is that of the dignity and privacy of individuals (Al-kfairy et al., 2024). AI relies on the collection and analysis of vast amounts of data, with potential violations of the privacy of those whose data is used. In an era of increasing surveillance and tracking of user activity, there is a risk that AI will become a surveillance tool, violating the autonomy and privacy of individuals. This introduces questions about compliance with ethical standards of human dignity and how personal data should be protected. Organizations implementing AI must therefore balance the effectiveness of their solutions with respect for users' privacy (Afroogh et al., 2024).

Ethical dilemmas have a direct impact on the level of public trust in AI technology (Russell & Norvig, 2016). Public trust is crucial to the acceptance and success of technology, especially when the technology directly impacts people's lives. A lack of transparency and understanding of how AI works can lead to skepticism and distrust of the decisions it makes. For example, if AI is used in recruitment processes or the granting of credit, a lack of explanation of the rules of operation can lead to feelings of injustice and discrimination among those who experience the negative effects of its operation (Dirgová Luptáková et al., 2023). In addition, ethical issues, such as lack of accountability for AI decisions, can undermine users' confidence in AI-based systems. If users are unsure who is responsible for the results generated by AI, they may fear that they will have no way to assert their rights in the event of errors. In addition, concerns about invasion of individual privacy may lead users to be reluctant to use technologies that use their personal information. In an era of growing public awareness of privacy and data protection, the lack of appropriate ethical standards for AI's use of data may lead to a loss of public trust and block further development of the technology (Dirgová Luptáková et al., 2023).

In response to these challenges, many international organizations and institutions have taken steps to develop ethical frameworks for AI. These aim to minimize risks and promote responsible and transparent use of technology (Florida & Cowls, 2022). An example is the guidelines developed by the IEEE (Institute of Electrical and Electronics Engineers), which focus on the ethical design and implementation of AI. The IEEE has proposed a set of principles, such as attention to privacy, accountability, transparency, and respect for human dignity. These guidelines are intended to enable systems that operate by ethical principles, thereby contributing to public trust in AI (O'Grady et al., 2022). The European Union has also set the stage for ethical AI development with its Ethics Guidelines for Trustworthy and Transparent AI. The document sets three main principles that should guide AI systems: legality, ethics, and technical soundness (Jobin et al., 2019). The EU guidelines place particular emphasis on accountability, transparency, and data protection, noting the need to ensure that AI technologies are designed and implemented responsibly and legally. In addition, organizations such as the OECD (Organization for Economic Cooperation and Development) and AI Ethics Lab have developed their own AI ethics proposals, pointing to the importance of an ethical approach to technology development. The OECD notes the need for regulations that promote fairness, accountability, and transparency in the use of AI, to enhance the security and trustworthiness of these systems (Jobin et al., 2019). An ethical AI framework aims to enable socially responsible implementation of this technology and minimize the risks associated with its development and use (Al-kfairy et al., 2024). Although different organizations have adopted different approaches, the common goal is to ensure that AI technologies operate by ethical values and contribute to social trust. The implementation of such ethical principles becomes crucial for the sustainable development of AI, as these technologies will have an increasing impact on our lives and societies.

Impact of risks and ethical dilemmas on trust in AI

AI is playing an increasingly important role in the daily lives of societies, helping to automate processes, optimize resources, and improve efficiency in many sectors, including health care, finance, industry, and commerce (Russell & Norvig, 2016). However, public trust in AI technologies remains a fragile resource, vulnerable to the impact of numerous risks and ethical dilemmas (Dirgová Luptáková et al., 2023). It outlines how specific risks and ethical challenges affect public trust in AI, how transparency and explainability are crucial in the process of building that trust, and how moral dilemmas related to AI decisions can lead to its erosion. Trust is at the core of the technology-user relationship and is one of the key factors determining the widespread acceptance and use of AI. In the case of AI technology, trust is built primarily based not only on its effectiveness, reliability, and security but also on the ethical values of its developers and the organizations implementing it (Mentzas et al., 2024). The risks and failures of AI, however, can seriously erode this trust, especially when they involve high-impact decisions affecting the lives of individuals or entire communities. The unpredictability of AI's actions and the possibility of algorithmic errors are among the main risks that undermine trust. Examples include erroneous decisions made by recruitment systems that exclude certain candidates from the hiring process based on faulty data or algorithmic biases or credit scoring systems that incorrectly assess the creditworthiness of customers

(Brynjolfsson & McAfee, 2014). Such cases lead to a sense of injustice and can create public concern that AI is not reliable or fair enough. From the perspective of an individual who has been harmed by a flawed algorithmic decision, trust in the technology is eroded, often leading to a reluctance to continue using it. Long-term maintenance of trust requires that AI systems be regularly reviewed, tested, and adapted to dynamically changing conditions and societal needs. In cases where AI technologies fail, organizations must respond quickly, providing users with an explanation of the causes of errors and a corrective plan. For many people, contact with AI that leads to a negative experience can create not only distrust in a specific tool but also in AI technology as a whole. In this way, AI risks and failures have the potential not only to undermine trust in individual products but also to undermine public confidence in the continued development and deployment of the technology (Brynjolfsson & McAfee, 2014). Transparency and explainability are key elements that contribute to building and maintaining public trust in AI. Transparency means that the AI's decision-making processes are understandable and accessible to users and that users have insight into how the algorithm works and how it operates. Explainability, in turn, means that users can understand how the algorithm makes decisions and what factors influence the outcome. For systems that affect people's lives – for example, in medicine, finance, or justice – transparency and explainability provide a foundation on which to build trust. In practice, transparency can be difficult to achieve, especially when complex algorithms are involved. However, a lack of transparency can lead to a situation where users begin to view AI as an unethical and inaccessible tool, further widening the gap between society and technology (Dirgová Luptáková et al., 2023). Ensuring explainability in the operation of AI algorithms, especially in areas where decisions have a significant impact on people, is therefore essential to building long-term and stable public trust (Florida & Cowls, 2022). The moral dilemmas associated with AI operations are one of the main factors that can negatively affect public trust in these technologies. As AI gains autonomy, questions arise about the moral principles that should guide AI and how it should make decisions in morally difficult situations. An example is autonomous systems used in automotive applications, such as autonomous cars, which may face an ethical dilemma in the event of a collision situation (Mentzas et al., 2024). Decisions made by AI tools in life-threatening situations are highly complex, and their moral rightness is difficult to assess, raising controversy and questions about the ethics of AI operations. Similarly, moral dilemmas arise in areas where AI affects equality and social justice. If algorithms are designed in a way that favors some groups over others, as is the case with algorithmic discrimination, this can lead to a loss of public trust. An example is the use of AI in recruitment processes, where historical biases can be unconsciously transferred to the results of candidate selection, leading to discrimination based on gender, age, or background (Nazeer, 2024). Such actions, perceived as morally wrong, raise public concerns and can influence perceptions of AI as an unreliable and unethical technology. Moral dilemmas affect the public acceptance of AI technology, especially in cases where the operation of algorithms is perceived to contradict ethical values. For AI technology to be fully trusted, it must operate in a manner consistent with social and ethical norms, which requires not only appropriate programming but also the creation of a legal and ethical framework to govern its operation in morally questionable situations.

Summary and conclusions

While the development of AI technologies brings enormous benefits in many fields, it also involves significant risks and ethical dilemmas that require a responsible approach to design, regulation, and education. In the face of these challenges, it is essential to take concrete steps that minimize risks, promote transparency, and increase public trust in AI. The design of AI systems should be based on ethical principles that minimize risks to users, ensure transparency and protect values such as privacy, justice, and individual dignity. Ethical AI design is an approach that creates socially responsible technologies that take into account the impact that their operation may have on diverse groups in society. One of the key elements of ethical AI design is transparency. Algorithms should be designed in a way that allows users to understand how the system works and what factors influence the decisions they make (Nazeer, 2024). The design of AI algorithms should take into account issues of fairness and anti-discrimination, and AI systems should be designed to protect users' personal information and privacy. AI designers must provide mechanisms for users to raise complaints and appeal decisions made by the system. The ability of a human to analyze the AI's performance and correct any erroneous decisions increases the sense of fairness and transparency, which positively affects the level of trust in the technology. One important step is the introduction of standards and certification systems for AI technologies. Certifications can ensure that an AI system meets minimum privacy, security, and anti-discrimination requirements. Lawmakers should clearly define who is responsible for decisions made by AI algorithms. Introducing clear regulations in this regard will help prevent those affected from seeking justice. Given the increase in the amount of personal data used by AI, the law should clearly define the rules for processing this data and the requirements for its protection. For the long-term development of AI, it is also necessary to increase public awareness of the capabilities and limitations of this technology. As AI becomes more ubiquitous, both end-users and professionals who create or implement these technologies must have adequate knowledge of its operation, risks, and responsibilities associated with it. The ethical design of AI, the introduction of a legal framework, and the education of the public are three key areas of action that can support the responsible deployment of AI technologies. Transparency and fairness are key to building public trust, and clear regulations can protect users from unethical use of the technology. Education and training, on the other hand, prepare both professionals and the public to use AI in an informed manner, minimizing risks and contributing to the long-term, sustainable development of the technology.

All of these factors have a significant impact on public trust in AI technology. With algorithms making decisions based on incomplete or flawed data, users may lose trust in systems that are not fully transparent or do not allow for appeal of decisions. Public trust is crucial to the acceptance of AI technology, and its loss can limit the widespread use of AI and negatively affect the further development of the technology. The further development of AI poses the need for researchers and practitioners to create an appropriate ethical and regulatory framework to minimize risks and protect the rights of individuals. AI will play an increasingly important role in areas such as health, finance, the labor market, education, or security, and responsible implementation of these technologies will require a thoughtful approach to issues such as transparency, equality, and accountability. Future AI development should be based on ethical principles that include respect for privacy, anti-discrimination, and accountability for erroneous decisions made by autonomous systems (Mentzas et al., 2024). The regulatory framework should also include guidelines for transparency and explainability of AI operations so that users can understand how decisions are made and what factors influence the performance of algorithms. Implementation of these principles will be key to building stable and sustainable public trust in AI technology, enabling its widespread use in society.

It is important to conduct further research to gain a deeper understanding of the risks and ethical dilemmas associated with this technology. It is necessary to develop tools and methods to monitor algorithms for biases, assess transparency, and be able to make corrections in case of wrong decisions. In particular, research on the explainability of algorithms and the assessment of their social impact can provide the knowledge needed to create more accountable and ethical systems. Introducing procedures to monitor the quality and fairness of AI decisions, as well as increasing access to ethics training in AI design, can contribute to more responsible development of these technologies. In addition, practitioners and designers of AI systems must understand their role in creating algorithms that will have a direct impact on people's lives. The conclusion of this study is the need for continued collaboration between researchers, practitioners, legislators, and the public to create an appropriate framework for AI technology that will support the development of this technology responsibly and ethically.

Acknowledgment

This research is partly funded by University of Economics Ho Chi Minh City, Vietnam.

References

Afroogh, S., Akbari, A., Malone, E., Kargar, M., & Alambeigi, H. (2024). Trust in AI: Progress, Challenges, and Future Directions. *arXiv preprint arXiv:2403.14680.*

Al-kfairy, M., Mustafa, D., Kshetri, N., Insiew, M., & Alfandi, O. (2024, August). Ethical challenges and solutions of generative AI: An interdisciplinary perspective. *Informatics*, *11*(3), 58. MDPI.

Bostrom, N., & Yudkowsky, E. (2018). The ethics of artificial intelligence. In *Artificial intelligence safety and security* (pp. 57–69). Chapman and Hall/CRC.

Brynjolfsson, E., & McAfee, A. (2014). *The second machine age: Work, progress, and prosperity in a time of brilliant technologies*. WW Norton & company.

Dirgová Luptáková, I., Pospíchal, J., & Huraj, L. (2023). Beyond code and algorithms: Navigating ethical complexities in artificial intelligence. In *Proceedings of the computational methods in systems and software* (pp. 316–332). Springer Nature Switzerland.

Floridi, L., & Cowls, J. (2022). A unified framework of five principles for AI in society. In S. Carta (Ed.), *Machine learning and the city: Applications in architecture and urban design* (pp. 535–545). John Wiley & Sons.

Guan, H., Dong, L., & Zhao, A. (2022). Ethical risk factors and mechanisms in artificial intelligence decision making. *Behavioral Sciences*, *12*(9), 343.

Jobin, A., Ienca, M., & Vayena, E. (2019). The global landscape of AI ethics guidelines. *Nature Machine Intelligence*, *1*(9), 389–399.

Kochetova, J. Y. (2024). Ethical risks of artificial intelligence and prospects for joint decision-making in medicine. *Čelovek*, *35*(3), 96–106.

Mentzas, G., Fikardos, M., Lepenioti, K., & Apostolou, D. (2024). Exploring the landscape of trustworthy artificial intelligence: Status and challenges. *Intelligent Decision Technologies*, *18*(2), 837–854.

Nazeer, M. Y. (2024). Algorithmic conscience: An in-depth inquiry into ethical dilemmas in artificial intelligence. *International Journal of Research and Innovation in Social Science*, *8*(5), 725–732.

O'Grady, K. L., Harbour, S. D., Abballe, A. R., & Cohen, K. (2022, September). Trust, ethics, consciousness, and artificial intelligence. In *2022 IEEE/AIAA 41st Digital Avionics Systems Conference (DASC)* (pp. 1–9). IEEE, Portsmouth, VA, USA.

Polemi, N., Praça, I., Kioskli, K., & Bécue, A. (2024). Challenges and efforts in managing AI trustworthiness risks: A state of knowledge. *Frontiers in Big Data*, *7*, 1381163.

Russell, S. J., & Norvig, P. (2016). *Artificial intelligence: A modern approach*. Pearson.

Treacy, S. (2023). Mechanisms and constraints underpinning ethically aligned artificial intelligence systems: An exploration of key performance areas. *Human-Centric Intelligent Systems*, *3*(3), 189–196.

2 Trust in generative artificial intelligence

A systematic literature review

Xuan Tai Mai and Trang Nguyen

Introduction

Generative artificial intelligence (GenAI) represents a significant leap in artificial intelligence (AI) and has become prevalent in many aspects of daily life. Recent reports indicate that GenAI users in the US increased from 7.8 million in 2022 to over 100 million in 2024, reflecting a growth rate of nearly 900% (Lebow, 2024). This novel technology has gained prominence due to its ability to engage in seemingly human-like conversations and create new content through text, images, and other forms of media in ways previously unimaginable (Dwivedi et al., 2023). GenAI systems, such as ChatGPT, Google Gemini, Claude, DALL-E, and Midjourney, employ neural network-based algorithms capable of generating diverse and original content after being trained on massive datasets (Riemer & Peter, 2024). This disruptive technology is expected to reshape industries, providing enhanced capabilities and opportunities for individuals and organizations in years ahead.

Unsurprisingly, the adoption of GenAI, particularly exemplified by ChatGPT, has been growing exponentially across various fields in recent years. The adoption rate of GenAI in organizations worldwide surged from 33% in 2023 to 65% in 2024, according to report by McKinsey (2024). For instance, in finance, AI-enabled agents are transforming customer interactions, facilitating banking activities, and providing expert advice on financial portfolios (Payne & O'Brien, 2024). In health care, GenAI is utilized in clinical documentation, medical diagnosis, treatment, and prognosis (Moulaei et al., 2024). The tourism industry benefits from GenAI's ability to handle traveler inquiries, design personalized itineraries, and facilitate travel planning and bookings (Sigala et al., 2024). In education, GenAI assists teachers with task management, tutoring support, and academic advising (Almogren et al., 2024). Additionally, marketers regard GenAI as an effective tool for enhancing marketing activities (Kshetri et al., 2023). While GenAI is expected to have substantial and ongoing impact across various aspects, integrating it into everyday life presents significant challenges and risks that must be carefully considered and effectively mitigated (Salah et al., 2024).

DOI: 10.4324/9781003586937-3

Unlike the conventional computing systems that rely on predefined algorithms as a sequential logic, GenAI advancements like OpenAI's ChatGPT and Google's Gemini generate predictions and unprecedented outputs based on probabilistic models derived from expansive datasets. As a result, although GenAI is powerful and helpful, concerns persist regarding its accuracy, reliability, and veracity (Riemer & Peter, 2024). For instance, Kim et al. (2023a) noted that ChatGPT occasionally struggles with logical queries and can produce entirely erroneous 'fact'. Hou et al. (2024) highlighted that AI systems can harbor algorithmic biases, leading to unfair decision-making. Moreover, GenAI can generate harmful or inappropriate content and fabricate information, which poses dangers in certain contexts (Fui-Hoon Nah et al., 2023). Despite GenAI's broad application and impact, it remains unclear how content-related issues influence user behavior and to what extend users can trust GenAI-generated responses.

Nonetheless, studies highlighting the significance of human-AI trust are still in their nascent stages (Henrique & Santos Jr, 2024), especially concerning GenAI. To date, there is scant research that provides an overview of this specific area. The most comprehensive literature review we are aware of was conducted by Henrique and Santos Jr (2024), who examined 470 papers related to trust in AI general across various disciplines up to January 2023. AI has gained phenomenal momentum in recent years, with much attention focused on the development and application of GenAI (Wach et al., 2023). Given the rapid pace of research in GenAI, the present review offers a more GenAI-specific context that aims to stimulate future research on trust in AI and ultimately foster users' acceptance of this novel technology.

In this chapter, we present a systematic literature review that maps out the current state of research on how users interact with and establish trust in GenAI. This chapter aims to comprehensively examine the determinants and consequences of trust constructs within GenAI settings. Focused discussions specific to GenAI, rather than broad generalizations, provide deeper insights into trust-related concerns in different sectors. By compiling perspectives on GenAI solicited across fields, our contributions to the AI literature are twofold. First, we point out the relevant concepts of trust in GenAI that have been addressed in recent studies, highlighting primary trust constructs that can serve as a foundation for future research. Second, we uncover the profound insights into the inhibitors, enablers, and consequences of trust in GenAI. This literature review offers a comprehensive understanding of the human-GenAI trust for various stakeholders, including organizations, AI providers, and academics, addressing the issues and challenges involved in building trust in GenAI.

The remainder of the study is structured as follows. Section 'Trust and the concepts of trust in GenAI studies' provides an overview of the theoretical backgrounds on trust and GenAI research. This is followed by a discussion of the main findings in section 'Main findings'. Finally, sections 'Implications, limitations, and future study directions' and 'Conclusion' outline the proposed implications and limitations of the study, suggest avenues for future research, and present the conclusion.

Trust and the concepts of trust in GenAI studies

Trust is a complex concept examined by scholars across various disciplines, including psychology, economics, and marketing (Doney & Cannon, 1997). Chen et al. (2015) indicated that trust is an essential element in the relationship between the trusting and trusted parties, as it mitigates risk-related behavior in the presence of opportunism and uncertainty. Trust plays an increasingly vital role in the use of information systems (ISs), given the growing complexity of modern technology nowadays (Söllner et al., 2016). Mayer et al. (1995) defined trust as one party's willingness to be vulnerable to the actions of another party. It is considered a fundamental catalyst in many transactional relationships (Wang et al., 2015) and is particularly relevant in the context of autonomous agent interactions, including human-AI relationships.

Several studies in the literature have adopted 'trust-in-GenAI' as a primary concept (Ali et al., 2023; Baek & Kim, 2023; Chang & Park, 2024; Marimon et al., 2024; Payne & O'Brien, 2024). Following Shin (2021)'s definition of trust, Ali et al. (2023) and Baek and Kim (2023) conceptualized trust in AI agents as the users' belief that the services and outputs provided by AI are reliable, credible, and trustworthy. Similar to trust in the IS domain, Chang and Park (2024) identified three key factors that define trust in AI systems: competence, benevolence, and integrity. Trust is established gradually through multiple interactions with AI agents and depends on users' willingness to accept risk. This view of trust focuses on the reliability and credibility of AI agents' results.

ChatGPT, a generative AI model, attracted 100 million users globally within two months of its launch in November 2022 (Kim et al., 2023a). As a consequence, ChatGPT has become a focal point in GenAI research and is often used interchangeably with the term GenAI in this study. A significant portion of the literature specifically addresses trust in ChatGPT, which belongs to the family of GenAI (Almogren et al., 2024; Jo, 2023; Kim et al., 2024; Lai et al., 2024; Pham et al., 2024; Polyportis & Pahos, 2024; Solomovich & Abraham, 2024; Xu et al., 2024). Scholars argue that trust is a crucial service element for ChatGPT, influencing its adoption and continued use. Trust enhances users' confidence in and reliance on ChatGPT's ability to provide truthful, accurate, and timely information (Almogren et al., 2024; Pham et al., 2024; Xu et al., 2024). More context specifically, Lai et al. (2024) defined trust in ChatGPT within educational contexts as students' perception of its reliability and trustworthiness in supporting their assessments. In tourism, travelers concerned about ChatGPT's trustworthiness are likely to avoid using it for travel information and advice (Xu et al., 2024).

Unlike traditional AI which primarily analyses data and infers outcomes, GenAI can create content based on user prompts. In most cases, users engage with GenAI, such as ChatGPT, by providing context and prompts to perform certain tasks. Because GenAI is expected to generate accurate, appropriate results beyond users' existing knowledge, the trustworthiness of GenAI-generated content is a pivotal construct in GenAI trust research. Studies by Kim et al. (2023a) and Kim et al. (2023b) focused on trust in the information provided by ChatGPT in the travelers' decision-making process. Chang and Park (2024) examined consumer trust in product recommended by GenAI agents. Additionally, Amos and Zhang (2024) investigated consumers' responses to the trustworthiness of online reviews generated by GenAI. Taken collectively, these studies suggest that while users react differently to GenAI applications, trust remains a critical challenge for modern users who are relied on and burdened by GenAI advancements. It also can be postulated that the trustworthiness of GenAI-provided information is essential for user adoption. The following section details the antecedents and consequences of how trust in GenAI is perceived by users and ultimately affects their behavioral responses.

Main findings

Antecedents of trust in GenAI

Enablers

This study identified various enablers (factors that increase trust in GenAI) and inhibitors (factors that decrease trust in GenAI), as this concept is widely recognized in IS research (Balakrishnan et al., 2024). Regarding enabling factors, Solomovich and Abraham (2024) acknowledged that users' perceptions of ease of use can enhance trust in ChatGPT. The authors suggested that the more a GenAI system is perceived as hassle-free and easy to use, the more likely users are to trust it. Ali et al. (2023) proposed four primary elements essential for trust in GenAI: relevance, credibility, usefulness, and intelligence. Their key findings revealed that ChatGPT's personalized recommendation, when seen as relevant, credible, useful, and intelligent, significantly enhances trust. Similarly, Chang and Park (2024) argued that GenAI's technological performance can reduce users' information search and cognitive effort by providing highly accurate and relevant information, which subsequently builds user trust. As expected, the perceived performance of a GenAI system plays a critical role in establishing initial trust.

Furthermore, Chakraborty et al. (2024) explored the factors influencing trust in the context of online shopping platforms that employ GenAI chatbots. The authors suggested that online platforms should emphasize the quality of interaction and credibility of their GenAI-powered chatbots to foster user trust. Additionally, Pham et al. (2024) investigated triggers for trust in ChatGPT and found that anthropomorphic cues, such as perceived warmth, communication speed, and perceived competence, are pivotal in forming user trust toward ChatGPT. In the tourism setting, Xu et al. (2024) defined perceived value as a traveler's appraisal of ChatGPT's functional benefits and added value, such as its 24/7 availability for recommendations, answers to queries, and customized travel itineraries. These features improve travelers' perceptions of ChatGPT's capability to deliver reliable and professional outputs that meet their needs. The chapter also demonstrated that perceived value is a key driver that activates travelers' perceived trust in the information provided by ChatGPT.

Beyond technology-related capabilities, an assemblage of users' motivational and external social factors also plays a role in building trust in GenAI. Using the uses and gratifications theory, Baek and Kim (2023) examined user motivations for utilizing ChatGPT and how these motives contribute to trust formation. Among five primary motivations, namely, information seeking, task efficiency, personalization, social interaction, and playfulness, they found that task efficiency and personalization significantly influence trust in GenAI, while information seeking, social interaction, and playfulness had no impact. Concerning external environmental factors, Xu et al. (2024) highlighted the importance of group norms and social approval in positively shaping users' perceived trust. These social influences function as accepted or implied regulations among group members, subsequently affecting individual trust in GenAI.

Based on the review of the relevant literature, it can be concluded that users' trust in GenAI is dynamic, primarily stemming from GenAI's technical capabilities as key determinants of trust. Additionally, internal motivational drivers and external social factors significantly contribute to the development of trust. To provide a more holistic understanding of the antecedents of trust, we further review the key inhibitors of trust in GenAI in the current literature.

Inhibitors

The review of existing literature reveals a notable gap in understanding the inhibiting factors of trust in GenAI. Through four different experiments, Kim et al. (2023a) examined the relationship between the quality of a GenAI chatbot's recommendations and users' trust. Their study concluded that the perceived trustworthiness of ChatGPT's recommendations decreased significantly when either the unethical or poor quality issues were salient. Similarly, Kim et al. (2023b) investigated the impact of various types of incorrect information generated by ChatGPT on travelers' decision-making processes. Their findings highlighted the detrimental effects of ChatGPT-generated incorrect information on users' trust. Drawing on status quo bias theory, Chakraborty et al. (2024) identified psychological factors, such as perceived inertia and perceived threat, as significant inhibitors of trust when using GenAI chatbots for online grocery shopping. Therefore, reducing perceived threats and inertia is crucial to fostering higher levels of user trust.

Impacts of trust in AI

The question of how trust in GenAI influences users' responses and behaviors is of significant interest to scholars across various fields. One of the most prominent impacts of trust in AI is its ability to drive users' behavioral intentions directly. Researchers often define behavioral intention as users' projected future actions concerning GenAI, including intentions to continue using the technology, share positive experiences, or recommend it to others. Studies consistently indicate that trust has a significantly positive impact on users' behavioral intentions, and this finding holds true across diverse areas such as tourism (Ali et al., 2023) and education (Jo, 2023; Lai et al., 2024). Additionally, many studies have specifically explored the direct effects of trust in GenAI on users' satisfaction (Pham et al., 2024), attitude (Polyportis & Pahos, 2024), purchase intention (Kim et al., 2024) (Chang & Park, 2024), usage intention (Payne & O'Brien, 2024), continuance intention (Baek & Kim, 2023; Pham et al., 2024), and word-of-mouth (Kim et al., 2024). Overall, these findings suggest that users' behavioral intentions can be aroused by their trust in GenAI.

Further, past research has unlocked the varying roles of trust in GenAI in different settings. For example, Solomovich and Abraham (2024) highlighted that trust in ChatGPT is associated with perceived usefulness of AI systems and mediates the relationship between perceived ease of use and perceived usefulness within the technological acceptance model. Conversely, Almogren et al. (2024) found that trust in ChatGPT does not significantly influence perceived usefulness in the education field, implying that while users may trust ChatGPT, they might not necessarily perceive it as particularly useful for their educational needs. Supporting this view, Lai et al. (2024) demonstrated that trust acts as a direct driver for the adoption of ChatGPT by students for assessment support. Similarly, trust was found to positively influence students' attitudes toward ChatGPT, which, in turn, increased their intention to use ChatGPT for educational purposes and engage in its actual use (Polyportis & Pahos, 2024).

In the tourism context, Xu et al. (2024) found that perceived trust in ChatGPT is a crucial factor that affects travelers' parasocial interactions with ChatGPT and their willingness to use it for travel-related services. Additionally, Kim et al. (2023a) indicated that the perceived trustworthiness of ChatGPT is a key determinant of travelers' acceptance and satisfaction with information recommended by ChatGPT. In workplace settings, the literature suggests that trust in GenAI is essential for fostering work engagement among employees and ultimately enhancing employee performance (Marimon et al., 2024).

In summary, findings related to GenAI highlight diverse user behavioral responses and reactions toward the usage of GenAI. While users' adoption and continuance intention are prominent themes in the current literature, other areas, such as purchase intention and satisfaction, are equally important and warrant further investigation. The contexts in which users adopt these novel technologies should also be considered, as outcomes may vary across different industries and applications.

Implications, limitations, and future study directions

Theoretical and managerial implications

With the growing importance of studying GenAI usage from the user's perspective, our review makes several contributions. First, to the best of our knowledge, this is the first comprehensive study to examine the antecedents and consequences of trust in GenAI. By conducting this systematic literature review, we advance new knowledge significantly by offering a thorough summary and mapping of the expanding body of research on trust in GenAI. This approach has allowed us to clarify established conclusions, highlight key insights, and identify gaps in the current complex literature on GenAI. Our review distinguishes key antecedents of trust in GenAI, encompassing both inhibitors and enablers, and emphasizes the need for further exploration of the consequences of trust in GenAI, as existing studies remain limited. This chapter deepens the discussion on users' decision-making processes related to GenAI and contributes more broadly to the AI literature.

Furthermore, our critical analysis and synthesis have bridged gaps in existing research through empirical insights drawn from diverse scientific domains and contexts. The findings have not only enhanced our understanding of the existing results but also framed future research directions that will strengthen the conceptual foundation of the topic of AI trust. We encourage future studies to integrate theoretical and conceptual insights from various disciplines to further develop this research domain.

We are also excited by the opportunities this chapter opens up from a managerial perspective, particularly in evolving existing or developing entirely new GenAI applications. Our study revealed that GenAI service providers systematically assess their offerings by examining the dual-factor mechanism of inhibitors and enablers that influence consumer adoption and engagement. The current literature suggests that the impact of GenAI can be better understood by considering the key drivers that foster trust in its usage, factors that support its implementation in workplaces or businesses, and outcomes that highlight the positive or negative effects of GenAI deployment. By uncovering these insights, we provide valuable implications for GenAI designers and developers aiming to enhance human interactions with GenAI applications. We offer strategies to build and nurture trust, ultimately leading to greater user engagement. Service managers can leverage these research findings to identify the critical determinants of trust and develop targeted initiatives that enhance user confidence and trust in GenAI.

Research limitations and future research

The findings we have presented in this chapter provides numerous opportunities for future investigation. First, our findings may be constrained by the limited scope of the literature reviewed. Specifically, the analysis was based on a relatively small number of articles due to its focus on GenAI settings. Future reviews could broaden their scope to explore trust mechanisms in AI more generally, offering a deeper understanding of the dynamics of trust. Additionally, future research should aim to review and develop frameworks that integrate both enablers and inhibitors within the broader context of AI usage to enrich its presence.

Moreover, our samples primarily focused on business, tourism, and education contexts. Future research could extend this by exploring other sectors that are intensively applying GenAI, such as health care (Moulaei et al., 2024). Lastly, with GenAI yet to evidence a series of developments, longitudinal studies are essential to examine the long-term effects of trust on users' behavioral responses. Such studies could provide insights into how trust in GenAI develops and changes over time, contributing to a more comprehensive understanding of its impact.

Overall, our study found that the use, challenges, and dynamics of trust in GenAI remain largely unexplored. We hope that this timely and thorough review will serve as a catalyst for innovative research in this field, drawing the interest of scholars and practitioners alike to enhance the available knowledge in the area of AI.

Conclusion

GenAI models, such as ChatGPT, Midjourney, and DeepBrain, are increasingly becoming the most disruptive technological breakthroughs in recent years and continue to hold immense potential for a wide range of applications across industries (Fui-Hoon Nah et al., 2023). Our study is grounded in the surprising insight that GenAI, due to its fundamentally different technical traits, often fails to meet conventional expectations of computing in terms of reliability, accuracy and veracity. We argue that building trust is one of the imperative strategies for GenAI to be successful. A lack of a coherent understanding of how to establish user trust in GenAI technologies contributes to the slow adoption of this technology.

In this chapter, we present a comprehensive review and critical analysis of the current research on trust in GenAI applications. We conducted an up-to-date examination of the emerging constructs of trust in GenAI, identifying its key drivers and outcomes. Furthermore, this chapter introduces and clarifies a dual pathway framework of factors that either enable or inhibit trust in GenAI. Enablers promote the users' trust in GenAI, while inhibitors help explain why users may hesitate to trust these systems. The findings provide a fresh perspective for researchers studying trust in GenAI and underscore areas for future research. We believe that this chapter will offer valuable insights for AI designers and developers seeking to enhance human interactions with AI systems and contribute to a deeper theoretical perspective of this complex subject (Table 2.1).

Acknowledgment

This research is partly funded by University of Economics Ho Chi Minh City, Vietnam.

Appendix

Table 2.1 Extant studies on trust in generative AI

Authors	*Construct of trust*	*Determinants of trust*	*Consequence of trust*	*Context, research methods*
Solomovich and Abraham (2024)	Trust in ChatGPT	Perceived ease of use	Perceived usefulness Behavioral intention	Tourism, survey ($N = 305$)
Ali et al. (2023)	Trust in AI agents	Relevance Credibility Usefulness Intelligence	Behavioral intention	Tourism, survey, ($N = 344$)
Kim et al. (2023a)	Perceived trustworthiness of ChatGPT's recommendations	ChatGPT's recommendation quality concerns ChatGPT's recommendation ethical concerns	Travelers' acceptance of and satisfaction with ChatGPT's recommendation	Tourism, 4 experiments
Baek and Kim (2023)	Trust in GenAI	Information seeking Task efficiency Personalization Social interaction Playfulness	Continuance intention	General ChatGPT users, survey ($N = 421$)
Jo (2023)	Trust in ChatGPT	N/A	Behavioral intention	ChatGPT users (students and office workers), survey ($N = 645$)
Kim et al. (2023b)	Perceived accuracy and trustworthiness of the information	Incorrect information	ChatGPT's recommendation acceptance and visit intention for recommended destinations	Tourism, six experiments
Chang and Park (2024)	Trust in AI recommender Trust in recommended products	Perceived performance	Intention to adopt consideration set	Retail, survey ($N = 443$)

Amos and Zhang (2024)	Review trustworthiness	Reviewer source (ChatGPT vs. human) Review authenticity	N/A	Retail, three experiments
Chakraborty et al. (2024)	Perceived trust toward online grocery shopping using GenAI chatbots	Perceived interaction quality Perceived credibility Perceived inertia Perceived theat Perceived regret avoidance Perceived attitude toward OGS using GenAI chatbots	N/A	Retail, survey ($N = 372$)
Almogren et al. (2024)	Trust in ChatGPT	N/A	Perceived usefulness Behavioral intention to use ChatGPT Acceptance of ChatGPT for Education	Education, survey ($N = 458$)
Lai et al. (2024)	Trust in ChatGPT	N/A	Perceived risk Behavioral intention to use, ChatGPT for assessment support	Education, survey ($N = 496$)
Payne and O'Brien (2024)	Trust in AI agents	N/A	Basic usage intention Advanced usage intention	Finance, survey ($N = 322$)
Xu et al. (2024)	Trust in ChatGPT	Social influence Perceived value	Parasocial interaction Intention to use	Tourism, survey ($N = 331$)
Marimon et al. (2024)	Trust in GenAI	User experience	Work engagement Employee performance	Workplace, survey ($N = 251$)
Polyportis and Pahos (2024)	Trust in ChatGPT	N/A	Student's attitude toward ChatGPT Behavioral intention Use behavior	Education, survey ($N = 355$)
Kim et al. (2024)	Trust in ChatGPT	N/A	Purchase intention Word of mouth	Tourism, survey ($N = 600$)
Pham et al. (2024)	Trust in ChatGPT	Perceived warmth Communication speed Perceived competence	Satisfaction Continuance usage intention	Tourism, survey ($N = 606$)

Source: Own research.

References

Ali, F., Yasar, B., Ali, L., & Dogan, S. (2023). Antecedents and consequences of travelers' trust towards personalized travel recommendations offered by ChatGPT. *International Journal of Hospitality Management, 114*, 103588.

Almogren, A. S., Al-Rahmi, W. M., & Dahri, N. A. (2024). Exploring factors influencing the acceptance of ChatGPT in higher education: A smart education perspective. *Heliyon, 10*(11), e31887.

Amos, C., & Zhang, L. (2024). Consumer reactions to perceived undisclosed ChatGPT usage in an online review context. *Telematics and Informatics, 93*, 102163.

Baek, T. H., & Kim, M. (2023). Is ChatGPT scary good? How user motivations affect creepiness and trust in generative artificial intelligence. *Telematics and Informatics, 83*, 102030.

Balakrishnan, J., Dwivedi, Y. K., Hughes, L., & Boy, F. (2024). Enablers and inhibitors of AI-powered voice assistants: A dual-factor approach by integrating the status quo bias and technology acceptance model. *Information Systems Frontiers, 26*, 921–942.

Chakraborty, D., Kar, A. K., Patre, S., & Gupta, S. (2024). Enhancing trust in online grocery shopping through generative AI chatbots. *Journal of Business Research, 180*, 114737.

Chang, W., & Park, J. (2024). A comparative study on the effect of ChatGPT recommendation and AI recommender systems on the formation of a consideration set. *Journal of Retailing and Consumer Services, 78*, 103743.

Chen, X., Huang, Q., Davison, R. M., & Hua, Z. (2015). What drives trust transfer? The moderating roles of seller-specific and general institutional mechanisms. *International Journal of Electronic Commerce, 20*(2), 261–289.

Doney, P. M., & Cannon, J. P. (1997). An examination of the nature of trust in buyer–seller relationships. *Journal of Marketing, 61*(2), 35–51.

Dwivedi, Y. K., Kshetri, N., Hughes, L., Slade, E. L., Jeyaraj, A., Kar, A. K., Baabdullah, A. M., Koohang, A., Raghavan, V., & Ahuja, M. (2023). Opinion paper: "So what if ChatGPT wrote it?" Multidisciplinary perspectives on opportunities, challenges and implications of generative conversational AI for research, practice and policy. *International Journal of Information Management, 71*, 102642.

Fui-Hoon Nah, F., Zheng, R., Cai, J., Siau, K., & Chen, L. (2023). Generative AI and ChatGPT: Applications, challenges, and AI-human collaboration, *Journal of Information Technology Case and Application Research, 25*(3), 277–304.

Henrique, B. M., & Santos Jr, E. (2024). Trust in artificial intelligence: Literature review and main path analysis. *Computers in Human Behavior: Artificial Humans, 2*(1), 100043.

Hou, T.-Y., Tseng, Y.-C., & Yuan, C. W. T. (2024). Is this AI sexist? The effects of a biased AI's anthropomorphic appearance and explainability on users' bias perceptions and trust. *International Journal of Information Management, 76*, 102775.

Jo, H. (2023). Understanding AI tool engagement: A study of ChatGPT usage and word-of-mouth among university students and office workers. *Telematics and Informatics, 85*, 102067.

Kim, J. H., Kim, J., Kim, C., & Kim, S. (2023a). Do you trust ChatGPTs? Effects of the ethical and quality issues of generative AI on travel decisions. *Journal of Travel & Tourism Marketing, 40*(9), 779–801.

Kim, J. H., Kim, J., Park, J., Kim, C., Jhang, J., & King, B. (2023b). When ChatGPT gives incorrect answers: The impact of inaccurate information by generative AI on tourism decision-making. *Journal of Travel Research, 64*(1), 51–73.

Kim, M. J., Kang, S.-E., Hall, C. M., Kim, J. S., & Promsivapallop, P. (2024). Unveiling the impact of ChatGPT on travel consumer behaviour: Exploring trust, attribute, and sustainable-tourism action. *Current Issues in Tourism*, 1–6.

Kshetri, N., Dwivedi, Y. K., Davenport, T. H., & Panteli, N. (2023). Generative artificial intelligence in marketing: Applications, opportunities, challenges, and research agenda, *International Journal of Information Management*, *75*, 102716.

Lai, C. Y., Cheung, K. Y., Chan, C. S., & Law, K. K. (2024). Integrating the adapted UTAUT model with moral obligation, trust and perceived risk to predict ChatGPT adoption for assessment support: A survey with students. *Computers and Education: Artificial Intelligence*, *6*, 100246.

Lebow, S. (2024). Generative AI hits 100 million user milestone in US. *EMARKETER*. https://www.emarketer.com/content/generative-ai-hits-100-million-users-milestone-us

Marimon, F., Mas-Machuca, M., & Akhmedova, A. (2024). Trusting in generative AI: Catalyst for employee performance and engagement in the workplace. *International Journal of Human–Computer Interaction*, 1–16.

Mayer, R. C., Davis, J. H., & Schoorman, F. D. (1995). An integrative model of organizational trust. *Academy of Management Review*, *20*(3), 709–734.

McKinsey. (2024). The state of AI in early 2024: Gen AI adoption spikes and starts to generate value. https://www.mckinsey.com/capabilities/quantumblack/our-insights/the-state-of-ai

Moulaei, K., Yadegari, A., Baharestani, M., Farzanbakhsh, S., Sabet, B., & Afrash, M. R. (2024). Generative artificial intelligence in healthcare: A scoping review on benefits, challenges and applications. *International Journal of Medical Informatics*, *188*, 105474.

Payne, E. H. M., & O'Brien, C. A. (2024). The search for AI value: The role of complexity in human-AI engagement in the financial industry. *Computers in Human Behavior: Artificial Humans*, *2*(1), 100050.

Pham, H. C., Duong, C. D., & Nguyen, G. K. H. (2024). What drives tourists' continuance intention to use ChatGPT for travel services? A stimulus-organism-response perspective. *Journal of Retailing and Consumer Services*, *78*, 103758.

Polyportis, A., & Pahos, N. (2024). Understanding students' adoption of the ChatGPT chatbot in higher education: The role of anthropomorphism, trust, design novelty and institutional policy. *Behaviour & Information Technology*, 1–22.

Riemer, K., & Peter, S. (2024). Conceptualizing generative AI as style engines: Application archetypes and implications. *International Journal of Information Management*, *79*, 102824.

Salah, M., Abdelfattah, F., & Al Halbusi, H. (2024). The good, the bad, and the GPT: Reviewing the impact of generative artificial intelligence on psychology. *Current Opinion in Psychology*, *59*, 101872.

Shin, D. (2021). The effects of explainability and causability on perception, trust, and acceptance: Implications for explainable AI. *International Journal of Human-Computer Studies*, *146*, 102551.

Sigala, M., Ooi, K.-B., Tan, G. W.-H., Aw, E. C.-X., Buhalis, D., Cham, T.-H., Chen, M.-M., Dwivedi, Y. K., Gretzel, U., & Inversini, A. (2024). Understanding the impact of ChatGPT on tourism and hospitality: Trends, prospects and research agenda. *Journal of Hospitality and Tourism Management*, *60*, 384–390.

Söllner, M., Hoffmann, A., & Leimeister, J. M. (2016). Why different trust relationships matter for information systems users. *European Journal of Information Systems*, *25*(3), 274–287.

Solomovich, L., & Abraham, V. (2024). Exploring the influence of ChatGPT on tourism behavior using the technology acceptance model. *Tourism Review*, Vol. ahead-of-print.

Wach, K., Duong, C. D., Ejdys, J., Kazlauskaitė, R., Korzynski, P., Mazurek, G., Paliszkiewicz, J., & Ziemba, E. (2023). The dark side of generative artificial intelligence: A critical analysis of controversies and risks of ChatGPT. *Entrepreneurial Business and Economics Review*, *11*(2), 7–30.

Wang, S. W., Ngamsiriudom, W., & Hsieh, C.-H. (2015). Trust disposition, trust antecedents, trust, and behavioral intention. *The Service Industries Journal*, *35*(10), 555–572.

Xu, H., Law, R., Lovett, J., Luo, J. M., & Liu, L. (2024). Tourist acceptance of ChatGPT in travel services: The mediating role of parasocial interaction. *Journal of Travel & Tourism Marketing*, *41*(7), 955–972.

3 Trust in content published on social media

Agnieszka Werenowska and Ewa Jaska

Introduction

The growing popularity of social media (SM) justifies conducting analyses of user trust in content disseminated through SM platforms. Modern SM plays an important role in the distribution of information and marketing campaigns. Additionally, such activities are supported by the new generation of AI technologies, which are supposed to imitate human behavior. New technologies make everyday life easier and make expected content appear on screens, although there is a problem of trust and credibility. New generation technologies are supposed to imitate the competences of human marketers, but to be effective, they should inspire trust. The issue of trust placed in them is becoming increasingly problematic, and therefore, the main goal of this chapter is to recognize and present the SM types and content formats perceived by users as trustworthy. The following specific objectives were formulated for this study:

- to identify the SM most frequently used by Generation Z representatives,
- to identify the most trusted SM,
- to identify the source of the most trusted content, including promotional content,
- to identify the trustworthy format of user-generated content (UGC),
- to determine the extent to which content created by AI is perceived as credible and which formats are most trusted by Generation Z representatives.

It is particularly important to investigate the perceptions of Generation Z representatives as the main users of SM. The theoretical part of this chapter presents the literature review, and the empirical part presents the results of the survey.

Literature review

Defining social media

The use of SM has become widespread, especially among the younger generation. IAB Polska (2019) defines SM as platforms and applications that are used for

DOI: 10.4324/9781003586937-4

online communication, offer users to set up virtual profiles, and rely on UGC. Fabjaniak-Czerniak (2012) defines SM as all methods of transmitting information online, thanks to which people create communities, build relationships, or exchange various contents such as news, opinions, ideas, and views. SM is also understood as a group of Internet applications that originate in the ideological and technological foundations of Web 2.0. These are applications that allow users to create their own content and exchange it with other members of the online community (Kaplan & Haenlein, 2010).

The categorizations of SM vary but most often include social networking sites, discussion forums, media sharing platforms, consumer review networks, blogging and publishing networks, Wiki pages, or, the oldest of them, mailing lists. All these tools are derived from the Web 2.0 trend and their similar features group them under the umbrella term of SM. Content published on SM can take the form of entries (posts), photos, graphics or videos, and animations.

The most popular SM networks in the world in 2024 (Brendhart, 2024) included: Facebook (3.06 billion), YouTube (2.5 billion), Instagram (2 billion). In Poland, the top three include Facebook, Facebook Messenger, Instagram (Digital Poland, 2024). Facebook is a free social networking site enabling registered users to create groups and social networks and share messages, films, or photos. YouTube is equally popular among Generation Z representatives. It is a video hosting platform that enables users to stream video content. An important part is content monetization, a process for creators who meet the appropriate conditions, i.e. reach a required watch-time hours and publish content regularly. Also, Instagram is becoming increasingly popular as it combines the features of a media-sharing platform with a social platform. As it enables users to publish photos and videos and conduct live broadcasts, it is becoming a place where brands can visually present their products by creating engaging content. Instagram also offers the feature stories, which allows users to publish short content that automatically disappears within 24 hours. Stories are ideal for quick communication of current promotional campaigns or behind-the-scenes, which build a sense of closeness and authenticity of the brand (Kietzmann & Canhoto, 2020).

SM have become a special space for users, facilitating social interactions, providing entertainment, and enabling information exchange. Marketers use SM to engage their customers and form lasting relationships with them. Brand engagement is an important area of marketing research that has been explored in studies for many years (Barger et al., 2016; Hollebeek et al., 2019). Nowadays, most people have access to the Internet and spend a lot of time scrolling through SM, which have become modern social trendsetters (Hanaysha, 2022). Therefore, the problem of trust placed in the content published on SM becomes a fundamental issue.

Trust in social media content

In literature, trust has been defined by many authors representing almost all scientific disciplines. It was extensively explored by, among others, Wereda

(2015, p. 217), who defined trust as the belief that "the other party shares our values and norms and will act to the benefit of their partner or at least will not harm them". The subject of trust was also studied by Paliszkiewicz (2010, 2014, 2019), who indicates that trust is associated with concern for the honesty of both parties, and it is voluntary, it is a matter of choice as one cannot be forced to trust. Trust is also associated with action, it is dynamic, it grows, or is breached. Trust can be considered a "bridge" between past behaviors and the expected future. She also emphasizes that trust is interpersonal, voluntary, accompanied by uncertainty about the honesty of the parties involved, temporary, and based on experience.

In the context of marketing activity, building a credible communication space with customers has become crucial. Trust and perceived security in SM are important factors impacting, e.g. the users' decisions to enter into online transactions (Werenowska & Jaska, 2022). SM enable the rapid dissemination of information on a large scale. However, the involved risk includes spreading misinformation, fake news, and manipulative content, which lead to the erosion of public trust in SM content, as discussed in studies by Vicari and Komendatowa (2023), Posetti and Matthews (2018), among others.

Assessing the credibility of content has become a challenge for both users and SM platforms. Leite and Baptista (2021) point out that consumers will establish or strengthen bonds with brands that are promoted by well-known and credible influencers. Considering the use of artificial intelligence in content creation, researchers are increasingly addressing the issue of trust and credibility of AI-generated content. The complexity of the phenomenon of trust dynamics in the era of digital technology is discussed by, among others, Gerlich (2024). This is a new, important aspect in the context of the dynamic development of new technologies, including SM. According to Lukyanenko et al. (2022), trust is a key challenge in the implementation and use of artificial intelligence. It is becoming an important factor in overcoming uncertainty that affects the development and implementation of artificial intelligence. In their research, Korzyński et al. (2024) addressed this aspect. Uncertainty and ambiguity lead to great caution, skepticism, and distrust. Undoubtedly, the increasingly important role of AI-powered tools is an irreversible trend. They are used not only as generative models employed to create content (both text, graphics, and video) but also in SM analytics and directly in advertising (IAB, 2019).

Methodology

For this research, we applied a diagnostic survey method using the online questionnaire technique (computer-assisted web interview). The questionnaire was distributed via SM platforms and e-mail to reach a sample of active SM users. This method allowed for quick and effective data collection and ensured access to the questionnaire to a wide audience regardless of their geographical location. The research used convenient sampling (Etikan, 2016). To forward the link to the questionnaire, the snowball technique was used where existing study subjects recruited future subjects from their acquaintances (Jabłońska & Sobieraj, 2013). The questionnaire consisted of rating questions and single- or multiple-choice closed-ended questions. The rating questions included a 5-point Likert scale, which is a widely recognized rating scale used in multiple studies, e.g. by Chikkabagewadi et al. (2022). The subjects were asked, among others, to indicate the most trusted SM, sources of the most trusted content, including promotional content, most trusted format of UGC, determinants of trust in content generated by artificial intelligence based on examples of AI influencers. The survey was conducted in June-July 2024 and involved 215 subjects – representatives of Generation Z, who declared using SM such as: Facebook, Instagram, YouTube, or TikTok. The majority of respondents were women (69.1%), participants with higher education (48.6%), and secondary education (47.2%), residents of cities with population exceeding 200,000 (45.2%).

Research results and discussion

Trust in promotional and user-generated content

The perceived credibility of information is strongly related to the trust placed in the source; for example, Geels et al. (2024) investigated the perceived credibility of SM posts based on users' verification of source credentials. In our study, we asked respondents to assess the perceived credibility of SM on a 5-point Likert scale. The rating showed that YouTube was perceived as the most credible platform (3.1) followed by Instagram and Facebook (2.8). The credibility of X app (i.e. former Twitter) was rated lower (2.6), and the lowest rating was given to TikTok (2.4). The obtained results regarding the credibility of SM do not correspond to their popularity among the Gen Z representatives. However, from the perspective of the effectiveness of marketing strategies, it is worth noting that activities characterized by lower credibility can also be effective. The attitude of the potential recipient of promotional messages is important. Lai and Liu (2020), on the other hand, indicate that the credibility of content and engagement in SM directly impacts users' acceptance of placing products in mobile social networks, while content likeability is an antecedent of SM engagement and content credibility. Hsiao-Ting et al. (2022) analyzed the relationship between SM marketing strategies and building consumer trust. They showed that consumer participation in SM communities supports higher levels of consumer trust and ratings of the surface credibility of information provided by an F&B vendor, and this surface credibility helps to develop consumer trust. Trust in the vendor also had a positive influence on consumer satisfaction with F&B product offerings.

In order to determine the level of consumer trust in promotional materials for products and services posted on SM, the respondents were asked to rate their own trust in them. About 61.2% of the study subjects described their trust in promotional materials in SM as a medium, 23.5% of respondents indicated low trust in promotional materials, and 11.1% declared high trust. Only 4.2% of respondents indicated the response "I do not trust". The obtained results show that Generation Z representatives using SM have limited trust in promotional content. The respondents were also asked to rate the factors influencing trust in promotional content. The user trust appeared to be mostly determined by previous experiences with a given brand or company (4.3) and previously obtained information or opinions about the brand (4.1). The trust of respondents is also greatly influenced by providing sources of information in promotional materials (3.6) and referring to research confirming the attractiveness of a product or service (3.4). The use of numerical data (3.0) and the attitudes and persuasive methods adopted by sellers (2.8) have a smaller impact on trust in promotional materials. Promotional materials in which products or services are endorsed by famous people, e.g. actors, athletes, or influencers, are not trusted (2.5). The detailed distribution of responses is presented in Figure 3.1.

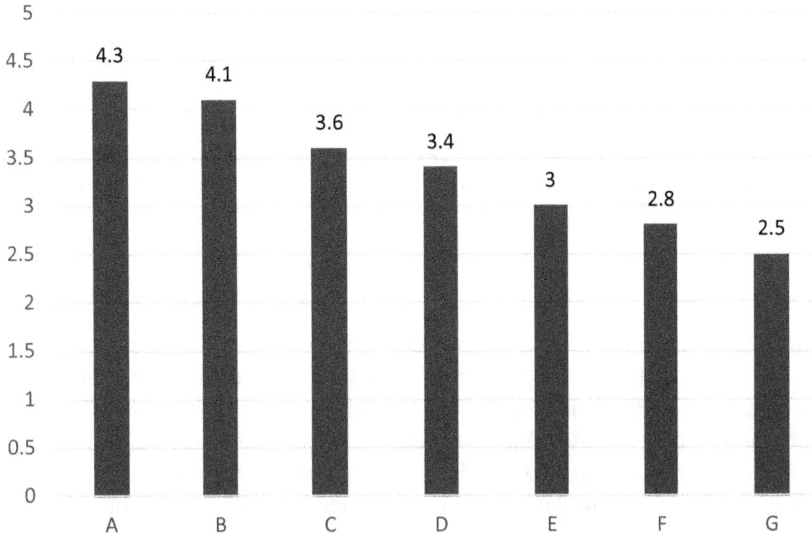

Figure 3.1 Factors influencing users' trust in promotional content in social media (scale 1–5)

(A) Previous experience with a given brand/company (good or bad); (B) previously obtained information or opinions about the brand; (C) providing sources of information; (D) reference to research on the attractiveness of the product or service; (E) use of numerical data; (F) sellers' attitudes and persuasive methods; (G) product endorsements by famous people, e.g. actors, athletes or influencers.

Source: Own research results.

McCroskey and Teven (1999) noted in their research that the credibility of content is perceived through the prism of the credibility of the source of information. In turn, Bratina and Faganel (2024) showed that the loyalty toward influencers is a constructive intermediary factor, amplifying the influence of credibility on consumer purchasing intentions. But consumer attitudes toward the brands endorsed by the influencers are modestly impacted by loyalty toward influencers. AbouElgheit (2024) analyzed the impact of promotional content generated by artificial intelligence and found that such content impacts purchase intentions of Gen Z representatives if they satisfy the need for entertainment, transparency, and usefulness.

In order to determine the most trusted format of UGC, we will refer to the results of a survey conducted in March-May 2024. The cited study involved 334 representatives of Generation Z. Most trustworthy UGC formats, as declared by the respondents, included reviews (4.23), Google reviews (4.1), and comments under posts (3.8). The respondents also assessed trust in the authors of content published on SM on a 5-point scale. The most trusted UGC authors included friends of SM users (4.04) and brand's customers posting comments and reviews (3.90). Positive customer reviews are usually more valuable than their number, as they reflect the real experiences of consumers in interactions with the company. Authentic, emotional reactions visible online can significantly influence the opinions of potential customers. In contrast, Gen Z consumers do not trust advertisements (2.12) or content generated by unknown authors (2.27) and influencers (2.55). Pradhan et al. (2022) emphasize, for example, that Generation Z tends to avoid products promoted by influencers when such communication is controlled by brands.

Trust in AI influencers

Considering that global AI-generated marketing revenues will reach over $107.5 billion by 2028, there is no doubt that AI will soon become an integral part of digital marketing (hubspot, 2024). The responses collected from our study subjects showed the extent to which AI-created content is perceived as trustworthy and which formats of such content inspire the greatest trust among Generation Z representatives. The SM users' trust in AI influencers was assessed on a 5-point Likert scale. The average trust rating for AI influencers was 2.37, which shows that Generation Z representatives do not trust AI influencers and have doubts regarding their content. Therefore, we asked the participants if AI influencers are the future of influencer marketing. Every fourth respondent believes that AI influencers may dominate the future of influencer marketing. This may be due to the perception of AI technology as innovative and capable of delivering personalized content in a scalable and cost-effective manner. However, 44% of respondents disagreed with this statement, which may indicate the current limitations of AI in terms of authenticity, emotional engagement, and creativity.

On the other hand, every third respondent had no opinion on this subject. This is probably because AI technology is still in the development phase, and many users are not yet sure how exactly its application will look in the future.

Consumer attitudes toward AI in the era of SM were the subject of studies, among others, by Sargın (2024). Data collected from 388 consumers in Turkey showed that a positive attitude toward AI has a significant positive impact on consumers' engagement in SM. Indecision may also result from a lack of sufficient knowledge about the possibilities and limitations of AI influencers. Schneider and Haried (2024), who conducted research among students, showed that despite the extensive use of AI, there are trust and anxiety concerns about the quality of outcomes and a lack of transparency in the knowledge acquisition process. Building trust in AI influencers requires further education, technology development, and adaptation to the needs of SM users.

The survey also aimed to identify the determinants of user trust in content disseminated by AI influencers. Every fourth respondent indicated the transparency of AI influencers' actions as a trust-building factor. Slightly fewer, i.e. 23% of respondents, expected evidence supporting AI influencers' recommendations, and 15% relied on positive reviews of other users. Content personalization was declared as the least important (7%).

The respondents were also asked whether they had ever noticed inconsistencies in content published by AI influencers; 17.5% noticed such irregularities, and 15.9% did not notice any inconsistencies. The majority, as many as 66.7% of respondents, did not pay attention to it. This is an important indicator that may affect users' trust in content generated by AI. Content inconsistencies may result from the technological limitations of AI and the lack of human supervision of the content creation process. Recurring content inconsistencies may lead to users' confusion and undermine their trust in AI influencers. It is worth noting the need to improve AI algorithms to minimize these discrepancies and increase message consistency. With the growing role of influencers in marketing, users expect the content they consume to be authentic and trustworthy.

The study subjects were asked to assess the impact of trustworthy content on trust in AI influencers on a 5-point scale, where 1 meant "strongly disagree" and 5 meant "strongly agree". The average score was 2.57, indicating a low impact of content authenticity on trust in AI influencers. The largest number of respondents rated this impact at 1, suggesting that content authenticity is not a key factor influencing trust. It seems that content authenticity, while important, does not appear to be a major determinant of SM users' trust in AI influencers. Users certainly expect truthful information, but other factors may have a greater impact on their trust. A lack of authenticity can lead to a certain loss of trust, but it is not the only or most important element affecting the credibility of AI influencers.

The results of the survey on trust in influencer recommendations show that the majority of respondents (62%) clearly prefer human influencers over AI-created influencers. A significant group of respondents (26%) do not have a clear preference, which may indicate openness to different forms of influence or lack of sufficient information to decide. These results suggest that human influencers are still valued more for their authenticity and ability to build trust with their audience. Almost every third respondent indicated a lack of authenticity and empathy as negative features (disadvantages) of AI influencers. The results of studies by Wiedmann and von Mettenheim (2020), Schouten et al. (2019), and Weismueller et al. (2020) indicate that people who perceive influencers as trustworthy and attractive experts are more likely to buy products promoted by them and support positive attitudes toward these products. Hussain and Zahid (2021) found that expertise has the greatest impact on a media influencer's credibility and the followers' purchase intentions, while the influencer's credibility has the greatest impact on the followers' attitude toward the brand.

Conclusions, limitations, and future research directions

The presented analysis indicates limited trust of Generation Z representatives in promotional content published on SM. Trust is most influenced by previous experiences with a given brand and acquired information or opinions about it. Respondents also demonstrated a relatively low level of trust in content generated by artificial intelligence and AI influencers.

We are aware that our research may have some limitations. These include limited research sample in terms of age, which cannot reflect general trends, subjectivity in the assessment of trust, rapidly changing nature of SM, narrowing the analysis to promotional content.

Future research directions should also be indicated: diversity of the studied population in terms of demographics and culture, multi-aspect analyses of the use of AI in marketing, misinformation, and fake news in content generated by AI.

References

AbouElgheit, E. (2024). The role of generative AI in shaping millennials and gen Z's orientation toward luxury products. *Journal of Marketing Development and Competitiveness*, *18*(2), 11–29. https://doi.org/10.33423/jmdc.v18i2.7012.

Barger, V., Peltier, J. W., Schultz, D. E. (2016). Social media and consumer engagement: A review and research agenda. *Journal of Research in Interactive Marketing, 10*(4), 268–287. https://doi.org/10.1108/JRIM-06-2016-0065.

Bratina, D., & Faganel, A. (2024). Understanding gen Z and gen X responses to influencer communications. *Administrative Sciences*, *14*(2), 33. https://doi.org/10.3390/admsci14020033.

Brendhart, G. (2024). Top 10 Most Popular Social Media Platforms, Retrieved from https://www.shopify.com/blog/most-popular-social-media-platforms (accessed May 14, 2024).

Chikkabagewadi, S., Devappa, V. M., & Karjinni, V. V. (2022). Evaluation of the bus service quality and a strategy for improvement service: Based on importance performance analysis and customer satisfaction index level. *Journal of Pharmaceutical Negative Results*, *13*, 5441–5453.

Digital Poland 2024, Retrieved May 14, 2024, from www.empemedia.pl/digital-poland-2024-raport-social-media-w-polsce.

Etikan, I. (2016). Comparison of convenience sampling and purposive sampling. *American Journal of Theoretical and Applied Statistics*, *5*(1), 1–4, https://doi.org/10.11648/j.ajtas.20160501.11.

Fabjaniak-Czerniak, K. (2012). Internetowe media społecznościowe jako narzędzie public relations [Online social media as a public relations tool]. In K. Kubiak (Ed.), *Zarządzanie w sytuacjach kryzysowych niepewności*. Warszawa: Wyższa Szkoła Promocji.

Geels, J., Graßl, P., Schraffenberger, H., Martin, T., & Kleemans, M. (2024). Virtual lab coats: The effects of verified source information on social media post credibility. *PLoS One*, *19*(5). https://doi.org/10.1371/journal.pone.0302323.

Gerlich, M. (2024). Exploring motivators for trust in the dichotomy of Human—AI trust dynamics. *Social Sciences, 13*(5), 251. https://doi.org/10.3390/socsci13050251.

Hanaysha, J. R. (2022). An examination of social media advertising features, brand engagement and purchase intention in the fast food industry. *British Food Journal, 124*(11), 4143–4160. https://doi.org/10.1108/BFJ-08-2021-0907.

Hollebeek, L. D., Srivastava, R. K., & Chen, T. (2019). SD logic–informed customer engagement: Integrative framework, revised fundamental propositions, and application to CRM. *Journal of the Academy of Marketing Science*, *47*(1), 161–185.

Hsiao-Ting Tseng, Shanmugam, M., Magalingam, P., Shahbazi, S., & Featherman, M. S. (2022). Managing enterprise social media to develop consumer trust. *British Food Journal*, *124*(12), 4626–4643. https://doi.org/10.1108/BFJ-11-2020-0995.

hubspot, 2024 AI Trends for Marketers, Retrieved May 14, 2024 from www.hubspot.com/hubfs/hubspot_AI_%20Trend_Report_2024.pdf.

Hussain, A., & Zahid, A. (2021). Examining the Impact of Social Media Influencer's Sredibility Dimensions on Consumer Behavior. Master's thesis, University of Gävle, Gävle, Sweden.

IAB (2019), www.raportstrategiczny.iab.org.pl/social-media/ (accessed May 14, 2024).

Jabłońska, K., & Sobieraj, A. (2013). Dobór próby badawczej czynnikiem sukcesu w prowadzonych badaniach empirycznych: Obronność [Research sample selection a factor for success in ongoing empirical research: Defense]. *Zeszyty Naukowe Wydziału Zarządzania i Dowodzenia Akademii, Obrony Narodowej*, *2*(6), 40–48.

Kaplan, A. M., & Haenlein, M. (2010). Users of the world, unite! The challenges and opportunities of SocialMedia. *Business Horizons*, *53*(1), 59–68. https://doi: 10.1016/J.BUSHOR.2009.09.003.

Kietzmann, J. H., & Canhoto, A. (2020). *Bending the frame: Instagram's influence on business and society*. Berlin: Springer Nature, 45–67.

Korzyński, P., Silva, S. C. e, Górska, A. M., & Mazurek, G. (2024). Trust in AI and top management support in generative-AI adoption. *Journal of Computer Information Systems*, 1–15. https://doi.org/10.1080/08874417.2024.2401986.

Lai, I. K. W., & Liu, Y. (2020). The effects of content likeability, content credibility, and social media engagement on users' acceptance of product placement in mobile social networks. *Journal of Theoretical and Applied Electronic Commerce Research*, *15*(3), 1–19. https://doi.org/10.4067/S0718-18762020000300102.

Leite, F. P., & Baptista, P. D. (2021). The effects of social media influencers' self-disclosure on behavioral intentions: The role of source credibility, parasocial relationships, and brand trust. *Journal of Marketing Theory and Practice, 30*, 295–311. https://doi.org/10.1080/10696679.2021.1935275.

Lukyanenko, R., Maass, W., & Storey, V. C. (2022). Trust in artificial intelligence: From a Foundational Trust Framework to emerging research opportunities. *Electron Markets, 32*, 1993–2020. https://doi.org/10.1007/s12525-022-00605-4.

Mccroskey, J. C., & Teven, J. J. (1999). Goodwill: A reexamination of the construct and its measurement. *Communication Monographs, 66*, 90–103.

Paliszkiewicz, J. (2010). Zarządzanie zaufaniem i dzielenie się wiedzą – przegląd literatury [Trust management and knowledge sharing – A literature review]. *Prace Naukowe / Akademia Ekonomiczna w Katowicach. Technologie wiedzy w zarządzaniu publicznym, 10*, 315–322.

Paliszkiewicz, J. (2014). *Zaufanie w zarządzaniu [Trust in management]*. Warszawa, PL: Wydawnictwo Naukowe PWN.

Paliszkiewicz, J. (2019). *Przywództwo, zaufanie i zarządzanie wiedzą w innowacyjnych przedsiębiorstwach [Leadership, trust and knowledge management in innovative enterprises]*. Warszawa, PL: CeDeWu.

Posetti, J., & Matthews, A. (2018). A short guide to the history of 'fake news' and disinformation. *International Center for Journalists*. Retrieved from https://www.icfj.org/sites/default/files/2018-07/A%20Short%20Guide%20to%20History%20 (accessed May 14, 2024).

Pradhan, D., Kuanr, A., Pahi, S., & Akram, M. S. (2022). Influencer marketing: When and why gen Z consumers avoid influencers and endorsed brands. *Psychology & Marketing; 2023, 40*, 27–47. https://dx.doi.org/10.1002/mar.21749.

Sargın, S. (2024). Antecedents and consequences of consumers' attitudes towards artificial intelligence in social media. *Business and Economics Research Journal, 15*(3), 229–256. https://doi.org/10.20409/berj.2024.443.

Schneider, S., & Haried, P. (2024). Use of artificial intelligence in higher education with particular reference to automated content generation, trust and anxiety. *Journal of Business and Educational Leadership, 14*(1), 4–21. Retrieved from https://www.proquest.com/scholarly-journals/use-artificial-intelligence-higher-education-with/docview/3068163399/se-2 (accessed May 14, 2024).

Schouten, A. P., Janssen, L., & Verspaget, M. (2019). Celebrity vs. influencer endorsements in advertising: The role of identification, credibility, and Product-Endorser fit. *International Journal of Advertising, 39*(2), 258–281. https://doi.org/10.1080/02650487.2019.1634898.

Vicari, R., & Komendatova, N. (2023). Systematic meta-analysis of research on AI tools to deal with misinformation on social media during natural and anthropogenic hazards and disasters. *Humanities & Social Sciences Communications, 10*(1), 332. https://doi.org/10.1057/s41599-023-01838-0.

Weismueller, J., Harrigan, P., Wang, S., & Soutar, G. N. (2020). Influencer endorsements: How advertising disclosure and source credibility affect consumer purchase intention on social media. *Australasian Marketing Journal, 28*(4), 160–170. https://doi.org/10.1016/j.ausmj.2020.03.002.

Wereda, W. (2015). Orientacja na zaufanie a budowanie relacji. In T. Borys, P. Rogala, & P. Skowron (Ed.), *Zrównoważony rozwój organizacji – odpowiedzialne zarządzanie [Sustainability of the organization – Responsible management], Prace Naukowe Uniwersytetu Ekonomicznego we Wrocławiu [Research Papers of Wrocław University of Economics]*, 376, 216–233. Wrocław: Wroclaw University of Economics.

Werenowska, A., & Jaska, E. (2022). Trust as an element of building brand relations with environmental groups. In J. Paliszkiewicz & K. Chen (Ed.), *Trust, Organizations and the Digital Economy: Theory and Practice*, 272–282. Routledge: Taylor and Francis.

Wiedmann, K., & Mettenheim, W. (2020). Attractiveness, trustworthiness and expertise—Social influencers' winning formula? *Journal of Product & Brand Management, 30*, 707–725.

4 Unlocking artificial intelligence for all

Navigating the complexities of the artificial intelligence digital divide

Mirjana Pejić-Bach and Josip Marić

Introduction

The rise of digital technologies has brought the issue of gasp in the usage of digital technologies to the forefront, highlighting disparities in access to and use of information and communication technologies (ICTs). The digital divide (DD) refers to the gap between those who can effectively utilize digital tools and those who cannot (Arendt, 2008). Early research focused on infrastructure and accessibility (Barzilai-Nahon, 2006; Van Dijk, 2006), while more recent studies assessed variables such as digital economy, education, and e-government initiatives (Vassilakopoulou & Hustad, 2023). The COVID-19 pandemic exacerbated the DD, impacting education, healthcare, and other essential services for those with limited digital access (Porumbescu, 2020; Liu et al., 2023). Disruptive technologies like artificial intelligence (AI) further complicate the divide. While AI offers solutions, such as personalized learning and improved healthcare, it also risks widening inequalities through the emerging algorithmic divide (Anis, 2023; Yu, 2020). AI's role in reshaping sectors like journalism, agriculture, and healthcare has both positive and negative implications (Samuel-Okon & Abejide, 2024). As digital advancements continue, addressing infrastructure gaps and skill disparities within the AI domain remains essential to ensure equitable access to the benefits of society (Banh & Strobel, 2023).

This chapter focuses on the adoption of AI in the context of DD, considering the role of demographic factors as well as some situational ones, such as subjective and social norms. Since the research on this topic is emerging, we are deploying the narrative review, focusing on the most important scientific research. The research goals that we pose are as follows: (i) RG1: determine areas in which research on AI and the DD is emerging; (ii) RG2: investigate the impact of demographic factors such as age, education, and gender influence the adoption of AI technologies; (iii) RG3: identify the key barriers to adoption of AI technologies.

This chapter is composed as follows. After this brief introduction, the second part of this chapter provides a literature review, while the third part provides an

DOI: 10.4324/9781003586937-5

outlook on the research's methodology. Results are presented in the fourth part, followed by a discussion and conclusion in the last part of this chapter.

Literature review

The DD refers to the gap between those who have access to and can effectively use digital technologies and those who cannot due to various reasons, such as a lower level of economic development or the barriers related to specific demographic groups, such as the elderly or minorities (Arendt, 2008; Yoon et al., 2020). The DD often reinforces the negative starting point, providing an example of the Mathew effect (Merton, 1973; Krumsvik, 2008), creating an even larger gap between individuals, organizations, or nations with regard to their initial position.

Initial studies on the DD focused on the infrastructure, accessibility, and cost-effectiveness of computers and Internet usage (Barzilai-Nahon, 2006; Van Dijk, 2006). Contemporary research assesses digital development through more advanced aspects, such as ICT application in firms, the progression of ICT education, and governmental ICT initiatives (Çilan et al., 2009; Jaković et al., 2021; Huňady et al., 2024; Mihajlović et al., 2023).

The COVID-19 pandemic exacerbated the DD, highlighting the disparities in access to technology and internet connectivity. The reliance on digital platforms for remote work, education, and healthcare during lockdowns disproportionately impacted individuals and communities with limited or no access (Porumbescu, 2020). Students without internet access struggled to participate in online learning, widening the educational achievement gap (Liu et al., 2023; Pejić Bach et al., 2023). Even more worrisome, individuals without reliable internet access faced barriers to accessing telehealth services, potentially delaying or limiting their healthcare access (Saeed & Masters, 2021).

AI has been defined as the ability of machines to simulate human intelligence, performing tasks that typically require human cognitive abilities such as reasoning, learning, and problem-solving (Korteling et al., 2021). The field encompasses various subdomains, including computer vision, natural language processing, virtual assistants, robotic process automation, and machine learning (Jha et al., 2021).

The emergence of AI offers promising solutions to help bridge DD and ensure more equitable access to benefits from digital technologies (Bentley et al., 2024). AI offers promising solutions to bridge the DD across various sectors. In education, AI can provide personalized learning experiences and reach underserved populations through interactive voice platforms in local languages (Anis, 2023). AI in journalism can improve information accessibility and audience engagement, although its impact is considered both positive and negative (Hassan & Albayari, 2022). In healthcare, AI enhances diagnostics and patient care, while in agriculture, it increases productivity through precision farming (Samuel-Okon &

Abejide, 2024). However, challenges persist, including infrastructure limitations, skills gaps, and cultural resistance (Samuel-Okon & Abejide, 2024). For example, the emerging algorithmic divide threatens to exacerbate inequalities, potentially leading to algorithmic deprivation, discrimination, and distortion (Yu, 2020).

The determinants of AI adoption from a DD perspective encompass various technological, organizational, and environmental factors. Access to and skills in using ICTs are crucial for AI literacy and adoption (Celik, 2023). Firm characteristics such as size and use of intangible assets, along with the adoption of complementary technologies like big data and cloud computing, significantly influence AI implementation (DeStefano et al., 2022; Pejić Bach et al., 2020). At the government level, factors including internet security, regulatory quality, government effectiveness, and globalization play critical roles in AI adoption (Wang et al., 2024a). Computational thinking and cognitive absorption also contribute to AI literacy (Celik, 2023). The impact of these determinants can vary in magnitude and direction, suggesting the need for targeted strategies to bridge the AI adoption divide (Wang et al., 2024a).

Methodology

The literature selection was conducted in many phases. Scopus was queried with the phrase "DD" AND "artificial intelligence" OR AI. The investigation was conducted from 2017 to 2024, which resulted in the initial selection of 188 publications. Further selection was implemented to identify the research that investigates the determinants of AI adoption in the context of the DD. This methodology resulted in the selection of 12 publications across several journals and conference proceedings.

Results

Table 4.1 provides a summary of 12 publications exploring the factors influencing AI adoption in the context of the DD. The studies span multiple countries and focus on sectors like healthcare, e-government, education, and general AI usage. Common positive determinants of AI adoption identified include higher education levels, younger age, digital literacy, trust in AI, and perceived usefulness. Negative factors often involve lower education levels, older age groups, digital anxiety, and concerns over data leakage and inaccurate information. Neutral factors are less frequently reported but can include demographic variables like gender and age in certain contexts.

Healthcare

Fritsch et al. (2022) conducted a cross-sectional study at a German tertiary referral hospital from December 2019 to February 2020 to investigate how patients and

Table 4.1 Summary of research investigating the determinants of AI adoption in the context of DD

Authors (year)	*Country/respondents/ method*	*Dependent variable/ method*	*Positive*	*Negative*	*Neutral*
Healthcare					
Fritsch et al. (2022)	Germany/452 patients/Survey	AI in healthcare/SEM	Younger, male, higher education level, and digital affinity	Elderly, women, lower education	–
Arellano Carmona et al. (2022)	USA/2,437 users/ survey	AI-based symptom checker/logistic regression	Higher education, Black, Latino, specialist care users	Primary care users	Female, elderly, Asian or Pacific, have insurance
Zhang et al. (2020)	USA/203 patients/ Survey	AI-diagnostics/ ANOVA, chi-square	Higher education and technology literacy	–	–
Wang et al. (2024b)	USA/508 patients ≥ 45 years/Survey	AI usage in outpatient service	Perceived usefulness, perceived ease of use, subjective norm	Digital anxiety	–
e-Government					
Jang (2024)	South Korea/48,420 citizens aged 14–90/Survey	AI in e-government/ SEM	Trust in AI, proactive privacy-protective behavior	Data leakage and inaccurate information	Online skills
Moreira and Naranjo-Zolotov (2023)	Portugal/208 citizens/ Survey	AI in e-government/ SEM	perceived usefulness, trust in AI, and social influence	–	–

(Continued)

Table 4.1 (Continued)

Authors (year)	*Country/respondents/ method*	*Dependent variable/ method*	*Positive*	*Negative*	*Neutral*
Education					
Tafazoli (2024)	Iran/24 language teachers/interviews and essays	Generative AI usage in teaching/qualitative	Accessibility, personalization, technology, overcoming isolation	–	–
Ramírez-Montoya et al. (2023)	Ecuador, Guatemala, and Mexico/115 educators/survey	Generative AI usage in teaching/descriptive statistics	Perception of AI as positive for tasks, e.g. automated assessment	Concerns for replacement of teachers, dependence on technology, ethics, privacy	
General usage					
Iancu and Iancu (2023)	Romania/235 48–78 aged citizens/ Survey	Chatbot acceptance/ SEM	Perceived ease of use, effort, competence, external control, perceived usefulness, social influence	–	Gender, age
Celik (2023)	Turkey/1,200 students/Survey	AI-based technologies in daily life/SEM	Computational thinking, access to ICTs, motivations, and skills to use ICTs	–	–
Park et al. (2022)	US/National survey	AI-based digital assistants/descriptive statistics	Higher income, higher education	Lower income, lower education	–
Woźniak-Jęchorek et al. (2023)	Poland/630 full-time working women	AI usage in work/ descriptive statistics	Higher education levels, larger cities, younger with less work experience	–	–

Source: Authors' work.

their companions perceive AI in healthcare. Among 452 participants, 53.18% viewed AI positively, while 4.77% had negative attitudes. Key determinants shaping these perceptions included age, gender, education level, and digital affinity. Older participants, women, and those with lower education or technical familiarity were more cautious about AI adoption, emphasizing the need for physician oversight in its application.

Arellano Carmona et al. (2022) conducted a cross-sectional study to examine the use of a web-based AI-powered symptom checker, Buoy, among health information seekers. This chapter, involving 2,437 users, assessed comprehensibility, confidence, usefulness, health-related anxiety, empowerment, and intention to use the tool in the future. Most users were well-educated, female, and White, with symptoms such as pain, gynecological issues, and masses accounting for over half of the visits. Users generally reported high confidence in Buoy, finding it useful and anxiety-reducing. Notably, Latino and Black users had significantly higher confidence and stronger intentions to discuss recommendations with a provider compared to White users.

Zhang et al. (2020) surveyed 203 participants to explore lay individuals' perceptions of AI-empowered healthcare systems, specifically focusing on AI's role in helping users comprehend diagnostic results. This chapter found that although awareness and experience with patient-facing AI systems were generally low, participants had a positive attitude toward these technologies. Many expressed comfort and willingness to use AI-driven health systems, believing AI could assist them in understanding diagnostic information. Key factors influencing perceptions included education background, technology literacy, and health literacy, with more educated and tech-savvy individuals showing higher trust in AI outputs.

Wang et al. (2024b) conducted a study to examine the factors influencing middle-aged and older individuals' acceptance of AI in outpatient services using the Technology Acceptance Model. Based on a survey of 508 participants, with 398 valid responses, this chapter found that facilitating conditions positively influenced perceived usefulness ($\beta = 0.665$, $P < 0.001$) and perceived ease of use ($\beta = 0.802$, $P < 0.001$), which in turn affected satisfaction. Digital anxiety negatively impacted attitudes toward using AI ($\beta = -0.243$, $P < 0.001$), while subjective norms positively influenced attitudes ($\beta = 0.706$, $P < 0.001$). This chapter highlights that perceived ease of use is critical for satisfaction, and future digital outpatient services need to address digital anxiety and subjective norms, as well as facilitate conditions to improve AI accessibility for middle-aged and older individuals, reducing DD in healthcare services.

e-Government

Jang (2024) conducted a study utilizing data from 48,420 South Korean respondents aged 14 to 90 to examine the determinants influencing AI usage, particularly in the context of e-government services. The research focused on individuals' online skills, trust in AI, and privacy-protective behaviors to assess their impact on adopting AI technologies. Findings indicated that trust in AI and proactive privacy-protective behaviors significantly and positively affect AI-based service usage, suggesting that when individuals feel capable of controlling vulnerabilities like data leakage and inaccurate information, they are more inclined to use AI services. Conversely, online skills did not have a significant effect on AI service adoption.

Moreira and Naranjo-Zolotov (2023) conducted an empirical study using an online survey with 208 respondents to explore the potential drivers behind citizens' acceptance of AI in e-government. This chapter found that perceived usefulness, trust in AI, and social influence are significant factors contributing to AI acceptance in e-government. While many participants were aware of AI and its potential in public administration, some needed to gain awareness of how AI could be effectively used in e-government.

Education

Tafazoli (2024) conducted a qualitative study to investigate the transformative potential of Generative Artificial Intelligence (GenAI), specifically ChatGPT, in addressing educational challenges faced by Iranian English language teachers. Through focus group sessions, semi-structured interviews, and reflective essays involving 23 teachers, this chapter identified five key determinants influencing AI adoption: accessible learning materials, personalized learning experiences, addressing ideological influences, overcoming technological barriers, and combating isolation from global trends. The findings highlight GenAI's capacity to provide diverse and up-to-date learning materials tailored to individual needs, address ideological biases, facilitate cross-cultural communication, and offer personalized professional development opportunities. By integrating GenAI into English language instruction, teachers can over come longstanding challenges, foster critical thinking, and promote intellectual freedom among students.

Ramírez-Montoya et al. (2023) conducted a study to analyze educators' perspectives on integrating generative AI into higher education teaching practices in Ecuador, Guatemala, and Mexico. Surveying 115 educators, the researchers found that teachers perceive AI-based tools as valuable resources for enhancing teaching and learning, especially appreciating features like automated assessment and student emotion detection. However, key determinants affecting AI adoption included challenges related to the DD and unequal access to technology, as well as concerns about potential teacher replacement, over-reliance on technology, ethical issues such as bias and data privacy, and academic integrity regarding the originality and authorship of AI-generated content.

General usage

Iancu and Iancu (2023) conducted a study to explore how middle-aged and aging adults in Romania perceive chatbots, particularly in the context of the COVID-19 pandemic. Using an online survey with 235 participants aged 40 to 78 and a covariance-based structural equation modeling approach, this chapter revealed that perceived ease of use, influenced by effort, competence, and external control, significantly impacts the perceived usefulness of chatbots. Moreover, perceived usefulness and social influence were found to drive behavioral intention to use chatbots, while gender and age had no significant effect.

Celik (2023) explored the determinants of AI literacy by developing a research model that includes DD, computational thinking, and cognitive absorption. This chapter found that computational thinking significantly enhances AI literacy, enabling individuals to use, recognize, and evaluate AI-based technologies like ChatGPT more effectively. Additionally, having physical access to ICTs increases the likelihood of individuals using and recognizing AI motivation and skills in using ICTs, which also contribute to better evaluation of AI outcomes and a more engaging experience with AI technologies. Given the impact of digital inequalities on AI literacy, future interventions should focus on addressing the four types of DD—motivational, usage, physical, and skills—to foster AI literacy effectively.

Park et al. (2022) investigated the DD in the access, use, and perceptual understanding of digital assistants, such as Google Home, Amazon Alexa, Echo, or Apple Siri, focusing on how socio-demographic factors shape inequalities in these areas. Analyzing two US national surveys, this chapter revealed that higher income individuals enjoy better access to and use of digital assistants, reflecting both first-level (access) and second-level (use) divides. Additionally, this chapter identified perceptual gaps based on education and demographics, with lower status individuals more likely to overlook risks like algorithmic bias, misjudgment, and data surveillance in favor of the convenience offered by AI-enabled devices.

Woźniak-Jęchorek et al. (2023) explored women's perspectives on integrating AI into professional work, focusing on their self-assessment of digital competencies. Based on a survey conducted among Polish women, this chapter found that AI adoption is influenced by factors such as age, education, work experience, and location. Women with higher education levels, residing in larger cities, and younger individuals with less work experience were more likely to recognize the benefits of AI. While these women were less concerned about the need to develop digital skills continuously, they expressed worries about AI's potential impact on work-life balance and earnings despite acknowledging its potential to enhance productivity and time management.

Discussion and conclusion

The swift growth of AI in the last several years indicates that it will have a tremendous impact on daily life, politics, the economy, and society in general. Therefore, decreasing the DD is of the utmost importance. The purpose of this chapter was to identify the research on AI adoption in the context of the DD, for which we conducted a review of papers extracted from the Scopus database. Since this research is emerging, we comprise the chapter as a narrative review. However, it is important to note that our research exclusively utilized the Scopus database as the source for studies addressing the AI DD. Moreover, other studies have addressed the factors of AI adoption without employing the phrase "DD" to describe the problem. Such studies were excluded from the sample, and only a restricted number of articles focusing solely on the AI DD were analyzed thoroughly.

Our research reveals that the research on the AI DD (RG1) is emerging across sectors like healthcare, e-government, education, and general AI usage, with studies conducted in countries such as Germany, the USA, South Korea, and Turkey. In healthcare, studies focus on AI-assisted diagnostics and patient acceptance of AI. E-government research examines how citizens adopt AI in public administration services, while education research explores generative AI's role in teaching, highlighting the importance of AI in personalized learning.

Demographic factors such as age, education, and gender significantly influence AI adoption (RG2). Younger individuals and those with higher education are more likely to adopt AI. Higher education positively affects AI usage in healthcare and e-government. Gender differences vary by context; in healthcare, males are more likely to adopt AI, while other studies find gender has no significant effect. Additionally, income and urban residency influence adoption, as higher access to technology promotes AI usage. Barriers to AI adoption were identified as digital inequality, anxiety about using AI, and concerns over privacy and ethics (RG3). Limited access to technology and digital skills are significant barriers where digital anxiety negatively impacts AI use in healthcare. Privacy concerns, particularly data leakage and ethical issues, are significant deterrents in both e-government and educational contexts. Additionally, mistrust of AI, driven by fears of data misuse or job displacement, reduces adoption rates, highlighting the need for greater transparency and user education.

References

Anis, M. (2023). Leveraging artificial intelligence for inclusive English language teaching: Strategies and implications for learner diversity. *Journal of Multidisciplinary Educational Research, 12*(6), 54–70. https://ijmer.s3.amazonaws.com/pdf/volume12/volume12-issue6(5)/9.pdf

Arellano Carmona, K., Chittamuru, D., Kravitz, R. L., Ramondt, S., & Ramírez, A. S. (2022). Health information seeking from an intelligent web-based symptom checker:

Cross-sectional questionnaire study. *Journal of Medical Internet Research, 24*(8), e36322. https://doi.org/10.2196/36322

Arendt, L. (2008). Barriers to ICT adoption in SMEs: How to bridge the digital divide? *Journal of Systems and Information Technology, 10*(2), 93–108. https://doi.org/10.1108/13287260810897738

Banh, L., & Strobel, G. (2023). Generative artificial intelligence. *Electronic Markets, 33*(1), 63. https://doi.org/10.1007/s12525-023-00680-1

Barzilai-Nahon, K. (2006). Gaps and bits: Conceptualizing measurements for digital divide/s. *The Information Society, 22*(5), 269–278. https://doi.org/10.1080/01972240600903953

Bentley, S. V., Naughtin, C. K., McGrath, M. J., Irons, J. L., & Cooper, P. S. (2024). The digital divide in action: How experiences of digital technology shape future relationships with artificial intelligence. *AI and Ethics*, 1–15. https://doi.org/10.1007/s43681-024-00452-3

Celik, I. (2023). Exploring the determinants of artificial intelligence (AI) literacy: Digital divide, computational thinking, cognitive absorption. *Telematics and Informatics, 83*, 102026. https://doi.org/10.1016/j.tele.2023.102026

Çilan, Ç. A., Bolat, B. A., & Coşkun, E. (2009). Analyzing digital divide within and between member and candidate countries of European Union. *Government Information Quarterly, 26*(1), 98–105. https://doi.org/10.1016/j.giq.2007.11.002

DeStefano, T. J., Teodorovicz, T., Cho, J., Kim, H., & Paik, J. (2022). What determines AI adoption?. *Academy of Management Proceedings, 2022*(1), 14791. https://doi.org/10.5465/AMBPP.2022.14791abstract

Fritsch, S. J., Blankenheim, A., Wahl, A., Hetfeld, P., Maassen, O., Deffge, S., ... & Bickenbach, J. (2022). Attitudes and perception of artificial intelligence in healthcare: A cross-sectional survey among patients. *Digital Health, 8*, 20552076221116772. https://doi.org/10.1177/20552076221116772

Hassan, A., & Albayari, A. (2022). The usage of artificial intelligence in journalism. In Hamdan, A., Harraf, A., Arora, P., Alareeni, B., & Khamis Hamdan, R. (eds) *Future of organizations and work after the 4th industrial revolution: The role of artificial intelligence, big data, automation, and robotics* (pp. 175–197). Cham: Springer International Publishing. https://doi.org/10.1007/978-3-030-99000-8_10

Huňady, J., Pisár, P., Khawaja, S., & Qureshi, F. H. (2024). The digital transformation of European Union countries before and during COVID-19. *Business Systems Research: International Journal of the Society for Advancing Innovation and Research in Economy, 15*(1), 22–45. https://doi.org/10.2478/bsrj-2024-0002

Iancu, I., & Iancu, B. (2023). Interacting with chatbots later in life: A technology acceptance perspective in COVID-19 pandemic situation. *Frontiers in Psychology, 13*, 1111003. https://doi.org/10.3389/fpsyg.2022.1111003

Jaković, B., Ćurlin, T., & Miloloža, I. (2021). Enterprise digital divide: Website e-commerce functionalities among European Union enterprises. *Business Systems Research: International Journal of the Society for Advancing Innovation and Research in Economy, 12*(1), 197–215. https://doi.org/10.2478/bsrj-2021-0013

Jang, C. (2024). Coping with vulnerability: The effect of trust in AI and privacy-protective behaviour on the use of AI-based services. *Behaviour & Information Technology, 43*(11), 2388–2400. https://doi.org/10.1080/0144929X.2023.2246590

Jha, N., Prashar, D., & Nagpal, A. (2021). Combining artificial intelligence with robotic process automation—an intelligent automation approach. In Ahmed, K. R., & Hassanien, A. E. (eds) *Deep learning and big data for intelligent transportation: Enabling technologies and future trends* (vol. 945, pp. 245–264). Cham: Springer. https://doi.org/10.1007/978-3-030-65661-4_12

Korteling, J. H., van de Boer-Visschedijk, G. C., Blankendaal, R. A., Boonekamp, R. C., & Eikelboom, A. R. (2021). Human-versus artificial intelligence. *Frontiers in Artificial Intelligence, 4*, 622364. https://doi.org/10.3389/frai.2021.622364

Krumsvik, R. J. (2008). From digital divides to digital inequality–The emerging digital inequality in the Norwegian Unitarian School. *Online Submission, 5*(9), 1–17.

Liu, Y., Hong, L., & Ma, G. (2023). Widening digital divide: Family investment, digital learning, and educational performance of Chinese High School students during the COVID-19 pandemic school closures. *Springer Science+Business Media, 18*(5), 2397–2413. https://doi.org/10.1007/s11482-023-10191-y

Merton, R. K. (1973). The Matthew effect in science. In *The sociology of science* (pp. 439–459). Chicago: Chicago University Press. (First published in Science, *159* (1968), 56–63).

Mihajlović, I., Stanković, M., & Djevojić, C. (2023). Digital skills towards competitiveness of human resources efficiency: Comparative approach. *ENTRENOVA-ENTerprise REsearch InNOVAtion, 9*(1), 29–44. https://doi.org/10.54820/entrenova-2023-0004

Moreira, J., & Naranjo-Zolotov, M. (2023). Exploring potential drivers of citizen's acceptance of artificial intelligence use in e-government. In *World conference on information systems and technologies* (pp. 336–345). Cham: Springer Nature Switzerland. https://doi.org/10.1007/978-3-031-45648-0_33

Park, Y. J., Lee, H., Jones-Jang, S. M., & Oh, Y. W. (2022). Digital assistants: Inequalities and social context of access, use, and perceptual understanding. *Poetics, 93*, 101689. https://doi.org/10.1016/j.poetic.2022.101689

Pejić Bach, M., Bertoncel, T., Meško, M., Suša Vugec, D., & Ivančić, L. (2020). Big data usage in European countries: Cluster analysis approach. *Data, 5*(1), 25. https://doi.org/10.3390/data5010025

Pejić Bach, M., Jaković, B., Jajić, I., & Meško, M. (2023). Investigating the impact of COVID-19 on e-learning: Country development and COVID-19 response. *Mathematics, 11*(6), 1520. https://doi.org/10.3390/math11061520

Porumbescu, G. (2020). The digital divide leaves millions at a disadvantage during the coronavirus pandemic. The Conversation. https://medicalxpress.com/news/2020-03-digital-millions-disadvantage-coronavirus-pandemic.html

Ramírez-Montoya, M. S., Oliva-Córdova, L. M., & Patiño, A. (2023). Training teaching personnel in incorporating generative artificial intelligence in higher education: A complex thinking approach. In *International conference on technological ecosystems for enhancing multiculturality* (pp. 163–175). Singapore: Springer Nature Singapore. https://doi.org/10.1007/978-981-97-1814-6_16

Saeed, S. A., & Masters, R. M. (2021). Disparities in health care and the digital divide. *Springer Science+Business Media, 23*(9). https://doi.org/10.1007/s11920-021-01274-4

Samuel-Okon, A. D., & Abejide, O. O. (2024). Bridging the digital divide: Exploring the role of artificial intelligence and automation in enhancing connectivity in developing nations. *Journal of Engineering Research and Reports, 26*(6), 165–177. https://doi.org/10.9734/jerr/2024/v26i61170

Tafazoli, D. (2024). Exploring the potential of generative AI in democratizing English language education. *Computers and Education: Artificial Intelligence, 7*, 100275. https://doi.org/10.1016/j.caeai.2024.100275

Van Dijk, J. A. (2006). Digital divide research, achievements and shortcomings. *Poetics, 34*(4-5), 221–235. https://doi.org/10.1016/j.poetic.2006.05.004

Vassilakopoulou, P., & Hustad, E. (2023). Bridging digital divides: A literature review and research agenda for information systems research. *Information Systems Frontiers, 25*(3), 955–969. https://doi.org/10.1007/s10796-020-10096-3

Wang, P., Mu, D., Xu, W., Zhang, X., Yang, X., Wang, C., & Yu, H. (2024b). Evaluating impact factors of middle-aged and older individuals' acceptance of artificial intelligence in outpatient service using the technology acceptance model. In *Knowledge organization for resilience in times of crisis: Challenges and opportunities* (pp. 249–260). Würzburg: Ergon-Verlag. https://doi.org/10.5771/9783987400476

Wang, S., Xiao, Y., & Liang, Z. (2024a). Exploring cross-national divide in government adoption of artificial intelligence: Insights from explainable artificial intelligence techniques. *Telematics and Informatics, 90*, 102134. https://doi.org/10.1016/j.tele.2024.102134

Woźniak-Jęchorek, B., Rydzak, W., & Kuźmar, S. (2023). Decoding the interface: Exploring women's perspectives on integrating AI in professional work. *Human Technology, 19*(3), 325–351. https://doi.org/10.14254/1795-6889.2023.19-3.2

Yoon, H., Jang, Y., Vaughan, P. W., & Garcia, M. (2020). Older adults' internet use for health information: Digital divide by race/ethnicity and socioeconomic status. *Journal of Applied Gerontology, 39*(1), 105–110. https://doi.org/10.1177/0733464818770772

Yu, P. K. (2020). The algorithmic divide and equality in the age of artificial intelligence. *Florida Law Review, 72*, 331.

Zhang, Z., Genc, Y., Xing, A., Wang, D., Fan, X., & Citardi, D. (2020). Lay individuals' perceptions of artificial intelligence (AI)-empowered healthcare systems. *Proceedings of the Association for Information Science and Technology, 57*(1), e326. https://doi.org/10.1002/pra2.326

Part II

Human-robot interaction, ethical considerations and trust

5 Trust in the context of psychological and managerial determinants of human-robot interaction

Joanna M. Moczydłowska, Agnieszka Tul-Krzyszczuk and Urszula Widelska

Introduction

The human-robot interaction (HRI) is a multidisciplinary area of research drawing from achievements in psychology, cognitive science, sociology, robotics, mechanics, electronics, and computer science, as well as management sciences and quality sciences. Studies in the field of HRI focus on the conditions of HRI and communication, as well as on understanding, designing, and assessing this interaction in order to best design and implement robots capable of performing interactive tasks in the human-operating environment, including work environment (Goodrich and Schultz, 2008, pp. 203–275; Rodriguez-Guerra et al., 2021, pp. 108557–108578, Gajšek et al., 2020, p. 6590). The complexity of HRI in terms of methodology is significant and does not come solely from the multi-disciplinary nature of research. It also results from the specificity of this interaction itself, where on the one hand, there is a complex technical artifact characterized by a specific intelligence, and on the other hand, there is a human who, as the subject of this interaction, perceives it, feels it, and has certain expectations toward it, usually higher than those toward own washing machine, car, or printer. Hence, the subject of research is not only the technical capabilities of robots but also the needs of users, their ideas, fears, expectations, and, in a broader perspective, also the ways of using robots in the context of lifestyles, models of societies, and questions about the patterns of the so-called good life (Małek-Orłowska, 2015, pp. 93–105; Scheutz et al., 2021; Dautenhahn, 2013). The specificity of the interaction between humans and robots directs researchers, especially in the era of Industry 4.0, to explore issues centered around these problematics in a broad context. Undoubtedly, a crucial issue is human trust in robots, considering the psychological context of the relationship and the managerial perspective.

Research in this area has been expanding in recent years, but there are still issues that require in-depth scientific exploration. The chapter uses the narrative literature review method. The following research question was formulated: What

DOI: 10.4324/9781003586937-7

determines people's attitudes toward robots? The obtained research results enable the identification of further directions of scientific research on HRI.

Trust as a component of human-robot interaction

Trust is an important element of relationships between humans and robots. It is a kind of expectation that the other party in the relationship will act in accordance with the values and rules required for a particular type of cooperation. It focuses on interactions that occur in teams consisting of humans and robots. From the viewpoint of management theory and practice, an extremely important element is the readiness of people to cooperate with robots while performing professional tasks (Goetz et al., 2003). Thanks to this solution, it is possible to combine the strengths of the machine, such as reliability, endurance, and repetition accuracy, with the strengths of humans, such as dexterity, flexibility, and the ability to make decisions. In the case of close cooperation between a human and a robot, the workspaces of the human and the robot overlap both spatially and temporally (Cohen et al., 2019, pp. 4037–4054). These robots can work side by side with humans on physically strenuous or monotonous tasks (Vido et al., 2020, pp. 1–21). Typical areas of application are "pick and place" tasks, object handling between different production stages, and "follow the line" applications, where the robot must precisely follow a designated movement path (e.g., for contour tracing or gluing work) (Kulik, 2015, pp. 102–103).

It should be noted that Robots perform their tasks in various social systems, and the analysis of HRIs in terms of the number of human and robot participants in a given team and the nature of the relationship between them (e.g., direction of information flow, autonomy) is the basis of the H. A. Yanco and J. L. Drury taxonomy (2004, pp. 2841–2846) covering 11 aspects of HRI, which clearly indicates the level of complexity of this interaction. Trust is a belief that defending common interests will be a priority for those constituting a given team and will facilitate the pursuit of shared goals (Mayer et al., 1995, p. 714). Placing a robot in this framework makes the bonds ambiguous – a robot is a machine perceived as an emotionless device serving humans. On the one hand, humans give it commands, expecting them to be carried out according to their will. On the other hand, however, the robot is a device, which means that its limited capabilities may not always fully meet human expectations. Due to its complexity, studying the human-robot relationship in the context of trust is of immense importance (Morze, 2023, p. 155). In the context of the human-robot relationship, trust involves a person's readiness to rely on the robot to perform tasks – activities that, in traditional forms of collaboration, are carried out by humans. Authority and tasks are delegated to the machine, with faith and conviction that it will perform them as well as, or even better than, a human (Baker et al., 2018, pp. 30–53). The machine ceases to be an object, a piece of equipment, and begins to be a collaborator, with collaboration based on trust (Oleson et al., 2011, p. 177).

Three groups of characteristics determining trust in the human-robot line can be distinguished (Morze, 2023, p. 149):

1 Characteristics specific to humans, e.g., skills, engagement, competences, specialized knowledge, workload, experience, prior, situational, awareness, demographic data, personality traits, attitude toward robots, comfort at work with robots, self-confidence, and inclination to trust.
2 Characteristics related to the environment, e.g., team, belongingness to a group, culture, communication, sharing of mental models, task types, task complexity, requirement, multitasking, and physical environment.
3 Characteristics specific to the robot, e.g., efficiency, properties, behavior, reliability, predictability, failure rate, level of automation, transparency, close presence, personality, adaptability, type, and anthropomorphism.

Other important aspects of trust in the human-robot relationship, identified by researchers, are as follows:

- Proximity between humans and robots – people tend to have greater trust in robots that coexist with humans, remaining in close proximity to them (Simon et al., 2020, p. 18). Human knowledge about the machine provides certainty in the relationship, based on the understanding that humans can responsibly control the machine (Ososky et al., 2014, p. 2). Individuals with higher levels of knowledge tend to trust their abilities, whereas those with less experience trust the robot (Sanders et al., 2011, p. 1432).
- Risk propensity – the less a person trusts the robot, the quicker they will intervene when it attempts to complete a task (Hancock et al., 2011, p. 518). On the other hand, excessive trust in the autonomy of robots can lead to the improper use of such systems when people overly rely on automation and fail to intervene in the event of critical failures (Chen et al., 2018, pp. 9–2).
- The system on which collaboration with the robot is based and the resulting level of work efficiency – the increase in work efficiency with the robot must correspond to total trust in the system (Setter et al., 2017, p. 1001; Arslan et al., 2021, p. 80).

According to M. Bergman et al. (2019), trust is the most important success factor in human-robot teamwork. The reliability and predictability of the robot can influence the level of trust. It is worth noting that confidence in a robot is only a human factor in this interaction, and the emotional states of robot operators may influence robot operation (Bauer et al., 2008, pp. 47–66). Therefore, a human-oriented perspective should also reflect the operator's contribution to the fluency of HRIs. Collaboration between a human and a robot can be implemented through the individual actions of each collaborator, including the human and the robot, and the operator's contribution supports the robotic system. Therefore, the description of the human-oriented perspective in HRI fluency by M. Paliga and A. Pollak (2021) is expanded to include the human performance of joint tasks with a robot, involvement, and contribution to HRI. This fluency perspective is referred to as human contribution-oriented. Research (Paliga, 2022, p. 104191) confirmed the significant impact of each perspective on the work efficiency of the team consisting of a robot and its operator.

Psychological determinants of the human-robot relationship and trust

Psychological determinants of the human-robot relationship and trust should be analyzed in terms of

- communication,
- attitudes, and
- propensity for cooperation.

One of the key challenges in HRI is effective communication (Peschel and Murphy, 2012, pp. 53–62; Cominelli et al., 2021, p. 9687). As we know, interpersonal communication is based, on the one hand, on verbal communication, i.e., on conducting conversations, but in many situations, it takes place using other means of expression. These are primarily gestures, changes in body position, static facial expressions, and dynamic changes of facial expressions (Bonarini, 2020, pp. 279–285). The way a human communicates with a robot depends on the type of signal used to transmit information (electrical, mechanical, sound, optical). It is also necessary to determine the method of producing this signal (generator) and its reception. Psychologists believe that most interactions in interpersonal communication are carried out non-verbally (Leathers, 2009). For this reason, robot designers try to implement both verbal and non-verbal communication mechanisms into humanoid robots. Such robots are equipped with the ability to interact with each other and implement social behaviors. In addition to the design of the robot's mechanical systems, which should ensure clear human-robot and robot-human communication and an appropriate sensory system for collecting information about the environment, the robot needs an efficient and effective control system that ensures dynamics adequate to the conditions of the task being performed (Randell et al., 2014, p. 52; Levine et al., 2018, pp. 421–436). Work in this area is very advanced (Raj and Kos, 2022). Due to the fact that emotions are an integral part of interpersonal interactions, significant efforts have been made to develop emotionally responsive robots. People can spontaneously form socio-emotional bonds with robots, even those that are not specifically designed to elicit social behaviors (Hoffman, 2019, pp. 209–218).

The key factor analyzed in the context of HRIs are attitudes toward robots. The concept of attitude is one of the most important concepts in social sciences (Grzegorczyk, 2017, pp. 34–44). Attitude is defined as "an individual predisposition to evaluate an object or a certain aspect of the world in a favourable or unfavourable way" (Robinette et al., 2017, pp. 129–155), which translates as liking or disliking products, services, people, thoughts, behavior, and other attitude items. Some authors add that it may be "durable" (Solomon, 2006, p. 242) or "relatively durable" (Wojciszke, 2011, p. 200). Such a definition emphasizes the importance of the affective (emotional-evaluative) aspect of attitude. It is also a factor constituting attitude, which may (but does not have to) also include other components: cognitive and behavioral. The behavioral aspect comes down to how we would like to behave toward the object of the attitude. However, the cognitive aspect of attitude concerns what we know and think about a given subject. Opinions should be treated as activation of attitudes consisting of evaluative and cognitive aspects. This means that the respondents' knowledge about robotics may significantly influence the constitutive, emotional-evaluative aspect of their attitudes (Robert et al., 2020, pp. 107–212). Therefore, the attitude toward robots is most often defined as a relatively persistent tendency to evaluate them positively or negatively. It can manifest itself through emotions, evaluations, automatic, and controlled reactions. Attitude allows to predict behavior toward robots, willingness to cooperate with them, and subjective or objective treatment, e.g., readiness to expose them to the risk of destruction (Różańska-Walczuk et al., 2016, pp. 15–24).

Research by A. Wasielewska and P. Łupkowski (2021, pp. 167–169) shows that there are many factors that influence HRIs, in particular people's attitudes toward robots. These include sex, education, age, culture, nationality, experiences related to robots, as well as the type of robots, the context in which they appear, the perceived usefulness and ease of use of robots, and finally the belief in uniqueness of human nature and the willingness to attribute positive human traits to robots. Beliefs about robots based on fictional characters from science fiction works and the image of robots created in the mass media also appear to be important (Bruckenberger et al., 2013, pp. 301–310), as well as the degree of similarity of robots to humans or broadly understood appearance of robots (Nomura et al., 2006, pp. 437–454).

The study by J. C. Giger et al. (2017, p. 509) allows the conclusion that an important element influencing attitudes toward robots is the belief in the uniqueness of human nature, as well as religiousness. A strong belief in the uniqueness of human nature is associated with negative emotions toward robots and, consequently, with more negative attitudes toward interactions with them. The same study also found that higher levels of religiousness are associated with more negative attitudes toward interacting with robots. Similar conclusions were formulated by M. Różańska-Walczuk et al. (2016, pp. 15–24). As a result of their research, the belief in the uniqueness of human nature appeared to be a significant predictor of attitudes toward robots with human characteristics and attitudes toward interactions with robots. The weaker the belief in the uniqueness of human nature, the more positive the attitude toward robots. Moreover, an interesting predictor of attitude toward robots was the willingness to attribute positive human characteristics, such as nice, pleasant, agile, and strong to them. The presented regularities apply to both adults and adolescents. In a study conducted on students in the last grades of primary schools, attitudes toward interactions with robots correlated with the belief in the uniqueness of human nature – these were positive relationships of moderate or low strength (Wasielewska and Łupkowski, 2021, p. 174). The variable which influences people's attitudes toward robots is the experience gained from contact with them. This concerns both real and virtual contact through various types of media and with fictional robot characters. Research results confirm that people who have ever seen a real robot in action or in the media had more positive attitudes toward interacting with them (Nomura, 2014, pp. 460–464; Łupkowski and Jański-Mały, 2020; Riek et al., 2011).

From the viewpoint of management theory and practice, an extremely important element is the readiness of people to cooperate with robots while performing professional tasks (Goetz et al., 2003). Thanks to this solution, it is possible to combine the strengths of the machine, such as reliability, endurance, and repetition accuracy, with the strengths of humans, such as dexterity, flexibility, and the ability to make decisions. In the case of close cooperation between a human and a robot, the workspaces of the human and the robot overlap both spatially and temporally (Cohen et al., 2019, pp. 4037–4054). These robots can work side by side with humans on physically strenuous or monotonous tasks (Vido et al., 2020, pp. 1–21). Typical areas of application are "pick and place" tasks, object handling between different production stages, and "follow the line" applications, where the robot must precisely follow a designated movement path (e.g., for contour tracing or gluing work) (Kulik, 2015, pp. 102–103).

Research provides evidence that people are generally willing to cooperate, for example, they are willing to give up some of their own particular interests if it contributes to achieving the collective interest (Wang et al., 2018, pp. 315–327). In the context of the subject of this chapter, an important question arises: does this readiness also apply to cooperation with robots? The answer to this question is very difficult. Robots perform their tasks in various social systems, and the analysis of HRIs in terms of the number of human and robot participants in a given team and the nature of the relationship between them (e.g., direction of information flow, autonomy) is the basis of the H.A. Yanco and J.L. Drury taxonomy (2004, pp. 2841–2846). The authors included 11 aspects of HRI, which clearly indicates the level of complexity of this interaction. These include task type, critical nature of the task, robot appearance (e.g., anthropomorphic – human-like robot, zoomorphic – animal-like one, and functional – machine-like robot), proportions between the number of people and the number of robots in the team, composition of the robot team, the level of coordination in the team, the role of a human in the interaction with a robot (the five roles: supervisor, operator, teammate, mechanic/programmer, and observer), a type of physical proximity between a human and a robot (interpersonal interactions, i.e., avoiding, passing, following, approaching, and touching), decision support for operators, space-time, level of autonomy/number of interventions (the number of human interventions necessary for the robot to work: from continuous intervention to full autonomy) (Yanco and Drury, 2004, pp. 2841–2846).

In the case of cooperation between humans and robots in manufacturing and production processes, interactions are most often analyzed with respect to four fluency perspectives: human emotion-oriented, human contribution-oriented, robot-oriented, and team-oriented (Paliga, 2022, p. 104191). The human-oriented fluency perspective of HRI is based on the assumption that people interact with robots as they would with other humans, not with tools. It reflects the cognitive and affective states of the human operator experienced while working with the robot. As such, it refers to a person's emotional state and is therefore referred to as a human emotions-oriented perspective. In particular, this perspective describes the level of confidence in the robot and the feeling that the robot commits to the team's success (Paliga and Pollak, 2021, p. 102698). According to M. Bergman et al. (2019), trust is the most important success factor in human-robot teamwork. The reliability and predictability of the robot can influence the level of trust. It is worth noting that confidence in a robot is only a human factor in this interaction, and the emotional states of robot operators may influence robot operation (Bauer et al., 2008, pp. 47–66). Therefore, a human-oriented perspective should also reflect the operator's contribution to the fluency of HRIs. Collaboration between a human and a robot can be implemented through the individual actions of each collaborator, including the human and the robot, and the operator's contribution supports the robotic system. Therefore, the description of the human-oriented perspective in HRI fluency by M. Paliga and A. Pollak (2021) is expanded to include the human performance of joint tasks with a robot, involvement, and contribution to HRI. This fluency perspective is referred to as human contribution-oriented. Research (Paliga, 2022, p. 104191) confirmed the significant impact of each perspective on the work efficiency of the team consisting of a robot and its operator.

Managerial dimension of human-robot interaction and trust

Managers' trust in robots is the result of combining technical effectiveness, reliability, compliance with regulations, and the ability to flexibly adapt to business needs. Building this trust requires a comprehensive approach that takes into account both the technical and organizational aspects of implementing robots in the workplace, especially in the era of Industry 4.0, where a systemic approach to competency management in the organization is necessary (Prifti et al., 2017; Enke et al., 2018). This systemic approach involves professionally and holistically shaping the competency area in the organization, as well as harmonizing it with other enterprise management subsystems (Badurek, 2014, pp. 80–83). The requirements posed by Industry 4.0 necessitate a continuous, systematic, and advanced program for developing and improving personnel skills, as well as ensuring the transfer of utilitarian engineering and managerial knowledge to enterprises, especially in the field of process management. It also becomes necessary to develop models related to the economic effects of Industry 4.0, which will allow monitoring of the effectiveness of adopted solutions. Completely new technologies require new technical skills, as exemplified by advanced robotics (Schmidt et al., 2015). In the modern enterprise, managers manage not only people but also robots. They are not ordinary machines or devices; they are members of teams and determine the quality of cooperation. According to analyses by the Boston Consulting Group, robots currently perform only 10% of production tasks they could perform, but by 2025, the share of robots is expected to increase to 25%. Similar growth is documented by experts from the analytical firm Gartner, estimating that in the United States, over 30% of jobs will be automated by 2025 (Gartner Inc., 2017). Therefore, people should primarily develop the skill of managing machine work. There is also a new skill emerging: communicating with machines. Advanced HRI, or HRI, has become an indispensable asset in Industry 4.0 (Moczydłowska, 2023). In particular, the implementation of collaborative robots (cobots) has changed the way modern, intelligent factories operate. These robots support operators by working alongside them on common tasks (Lin and Lukodono, 2021, p. 5990).

Since cobots are intended to be more like collaborators than tools, smooth interaction between them and their operators is crucial for task execution and, consequently, for their high work efficiency and effectiveness (Paliga, 2022, p. 104191). The growing popularity of robots and robotics means that we are dealing with a constantly growing ecosystem of robots around us (Palomäki et al., 2018, pp. 3–4). This ecosystem includes both real robots, such as industrial robots, autonomous vehicles, cleaning robots, and assisting robots, as well as robots appearing in film productions, animations, games, or as virtual assistants (e.g., Google Assistant, Siri). Robots are, of course, part of a larger whole that creates the technological reality around us. At the same time, they are an exceptionally interesting element because, by design (to a greater or lesser extent), they are meant to reflect humans.

Conclusion

Trust in human-robot relationships is a complex issue. In its analysis, the psychological background must be taken into account. Human trust in robots results from the processes of communicating with the robot and their consequences, people's attitudes toward robots, and their propensity for cooperation. In the conditions of Industry 4.0, building trust in and managing robots is also a particular challenge for contemporary managers. The robot ceases to be a machine and begins to be a member of the team that the manager must manage. The effectiveness of this process will depend on the attitudes of the team, as well as on the methods of communication among members of a team that is heterogeneous in terms of structure.

References

Arslan, A., Cooper, C., Khan, Z., Golgeci, I., & Ali, Im. (2021). Artificial intelligence and human workers interaction at team level: a conceptual assessment of the challenges and potential. HRM strategies. *International Journal of Manpower*, 43(1), 75–88, DOI: 10.1108/IJM-01-2021-0052.

Badurek, J. (2014). Systemy ERP dla wytwórczości nowej generacji, Przedsiębiorstwo we współczesnej gospodarce — teoria i praktyka. *ACM Transactions on Interactive Intelligent Systems (TiiS)*, 2, 79–90. https://dl.acm.org/toc/tiis/2018/8/4.

Baker, A. L., Phillips, E. K., Ullman, D., & Keebler, J. R. (2018). Toward an understanding of trust repair in human-robot interaction: current research and future directions. *ACM Transactions on Interactive Intelligent Systems (TiiS)*, 8(4), 30–53, DOI: 10.1145/3181671.

Bauer, A., Wollherr, D., & Buss, M. (2008). Human–robot collaboration: a survey. *International Journal of Humanoid Robotics*, 5(01), 47–66.

Bergman, M., de Joode, E., de Geus, M., & Sturm, J. (2019). Human-cobot teams: exploring design principles and behaviour models to facilitate the understanding of non-verbal communication from cobots. In *Proceedings of the 3rd International Conference on Computer-Human Interaction Research and Applications (CHIRA 2019)*, SCITEPRESS – Science and Technology Publications, pp. 191–198, DOI: 10.5220/0008363201910198.

Bonarini, A. (2020). Communication in human-robot interaction. *Current Robotics Reports*, 1, 279–285.

Bruckenberger, U., Weiss, A., Mirnig, N., Strasser, E., Stadler, S., & Tscheligi, M. (2013). The good, the bad, the weird: Audience evaluation of a "real" robot in relation to science fiction and mass media. In *Social Robotics: 5th International Conference, ICSR 2013*, Bristol, UK, October 27–29, 2013, Proceedings 5, 301–310.

Chen, M., Nikolaidis, S., Soh, H., Hsu, D., & Srinivasa, S. (2018). Trust aware decision making form human-robot collaboration: model learning and planning. *ACM Transactions on Human-Robot Interaction*, 9(2), 9–23, DOI: 10.1145/3359616.

Cohen, Y., Naseraldin, H., Chaudhuri, A., & Pilati, F. (2019). Assembly systems in Industry 4.0 era: a road map to understand Assembly 4.0. *The International Journal of Advanced Manufacturing Technology*, 105, 4037–4054.

Cominelli, L., Feri, F., Garofalo, R., Giannetti, C., Meléndez-Jiménez, M. A., Greco, A., & Kirchkamp, O. (2021). Promises and trust in human–robot interaction. *Scientific Reports*, 11(1), 9687.

Dautenhahn, K. (2013). Human-robot interaction. In M. Soegaard, & R. F. Dam (red.), *The Encyclopedia of Human-Computer Interaction*, 2nd edn, Interaction Design Foundation, Aarhus.

Enke, J. i in. (2018). Industrie 4.0—competencies for a modern production system: curriculum for learning factories. *Procedia Manufacturing*, *23*, 267–272.

Gajšek, B., Stradovnik, S., & Hace, A. (2020). Sustainable move towards flexible, robotic, human-involving workplace. *Sustainability*, *12*(16), 6590.

Gartner Inc. (2017). Gartner Says By 2020, Artificial Intelligence Will Create More Jobs Than It Eliminates. https://www.gartner.com/en/newsroom/press-releases/2017-12-13-gartner-says-by-2020-artificial-intelligence-will-create-more-jobs-than-it-eliminates (07.10.2024).

Giger, J. C., Moura, D., Almeida, N., & Piçarra, N. (2017, May). Attitudes towards social robots: The role of gender, belief in human nature uniqueness, religiousness and interest in science fiction. In *II International Congress Interdisciplinarity in Social and Human Sciences 11th - 12th May 2017*. Research Centre for Spatial and Organizational Dynamics University of Algarve, Faro, Portugal.

Goetz, J., Kiesler, S., & Powers, A. (2003, November). Matching robot appearance and behavior to tasks to improve human-robot cooperation. In *W: The 12th IEEE International Workshop on Robot and Human Interactive Communication, 2003*. Proceedings. ROMAN 2003 (ss. 55–60), IEEE, Millbrae.

Goodrich, M. A., & Schultz, A. C. (2008). Human–robot interaction: a survey. *Foundations and Trends® in Human–Computer Interaction*, 1(3), 203–275.

Grzegorczyk, T. (2017). Społeczeństwo wobec rozwoju robotyki: rola badań opinii publicznej. *Społeczeństwo i Ekonomia*, 1(7), 34–44.

Hancock, P. A., Billings, D. R., & Schaefer, K. E. (2011). A meta-analysis of factors affecting trust in human-robot interaction. *Human Factors*, 53(5), 517–527, DOI: 10.1177/0018720811417254.

Hoffman, G. (2019). Evaluating fluency in human–robot collaboration. *IEEE Transactions on Human-Machine Systems*, 49(3), 209–218.

Kulik, J. (2015). Ramię w ramię z robotem. Jak współpraca robotów i ludzi ukształtuje robotykę przyszłości. *Automatyka*, 9, 102–103.

Leathers, D. G. (2009). *Komunikacja niewerbalna: zasady i zastosowania*. Wydawnictwo Naukowe PWN, Warszawa.

Levine, S., Pastor, P., Krizhevsky, A., Ibarz, J., & Quillen, D. (2018). Learning hand-eye coordination for robotic grasping with deep learning and large-scale data collection. *The International Journal of Robotics Research*, 37(4-5), 421–436.

Lin, C. J., & Lukodono, R. P. (2021). Sustainable human–robot collaboration based on human intention classification. *Sustainability*, 13(11), 5990.

Łupkowski, P., & Jański-Mały, F. (2020). The more you see me the more you like me: influencing the negative attitude towards interactions with robots. Journal of Automation Mobile Robotics and Intelligent Systems, 14(3), 10–17, DOI: 10.14313/JAMRIS/3-2020/27.

Małek-Orłowska, M. (2015). Koncepcja poszerzonej epistemologii a metodologia badań interakcji człowiek-robot. *Filo-Sofija*, 15(29), 93–105.

Mayer, R. C., Davis, J. H., & Schoorman, F. D. (1995). An integrative model or organization trust. *Academy of Management Review*, 3(20), 709–734, DOI: 10.2307/258792.

Moczydłowska, J. M. (2023). *Przemysł 4.0. Ludzie i technologie*, Difin, Warszawa.

Morze, M. (2023). Zaufanie – istotny czynnik w human–robot collaboration. *Optimum*, 1(111), 141–155.

Nomura, T. (2014). Influences of experiences of robots into negative attitudes toward robots. In *The 23rd IEEE International Symposium on Robot and Human Interactive Communication (IEEE RO-MAN 2014)*, 25–29 Aug. 2014, At Heriot-Watt University, Edinburgh, Scotland, DOI: 10.1109/RO-MAN32039.2014.

Nomura, T., Suzuki, T., Kanda, T., & Kato, K. (2006). Measurement of negative attitudes toward robots: interaction studies. *Social Behaviour and Communication in Biological and Artificial Systems*, 7(3), 437–454.

Oleson, K. E., Billings, D. R., Kocsis, V., Chen, J. Y. C., & Hancock, P. A. (2011). Antecedents of trust in human-robot collaboration. In *IEEE International Multi-Disciplinary Conference on Cognitive Methods in Situation Awareness and Decision Support (CogSIMA)*, Miami Beach, FL, USA, 2011, pp. 175–178, DOI: 10.1109/COGSIMA.2011.5753439.

Ososky, S., Sanders, T., Jentsch, F., Hancock, P., & Chen, J. Y. (2014, June). Determinants of system transparency and its influence on trust in and reliance on unmanned robotic systems. In *Proceeding of SPIE – the International Society of Optical Engineering*, Bellingham, WA, 90840E-1–90840E-12.

Paliga, M. (2022). Human–cobot interaction fluency and cobot operators' job performance. The mediating role of work engagement: a survey. *Robotics and Autonomous Systems*, 155, 104191.

Paliga, M., & Pollak, A. (2021). Development and validation of the fluency in human-robot interaction scale: a two-wave study on three perspectives of fluency. *International Journal of Human-Computer Studies*, 155, 102698.

Palomäki, J., Kunnari, A., Drosinou, M., Koverola, M., Lehtonen, N., Halonen, J., ... & Laakasuo, M. (2018). Evaluating the replicability of the uncanny valley effect. *Heliyon*, 4(11), 3–4.

Peschel, J. M., & Murphy, R. R. (2012). On the human–machine interaction of unmanned aerial system mission specialists. *IEEE Transactions on Human-Machine Systems*, 43(1), 53–62.

Prifti, L., Knigge, M., Kienegger, H., & Krcmar, H. (2017). A competency model for "Industrie 4.0" employees. In J. M. Leimeister, & W. Brenner (Hrsg.), *Proceedings der 13. Internationalen Tagung Wirtschaftsinformatik (WI 2017)*, St. Gallen, 46–60.

Raj, R., & Kos, A. (2022). A comprehensive study of mobile robot: history, developments, applications, and future research perspectives. *Applied Sciences*, 12(14), 6951.

Randell, R., Greenhalgh, J., Hindmarsh, J., Dowding, D., Jayne, D., Pearman, A., ... & Kotze, A. (2014). Integration of robotic surgery into routine practice and impacts on communication, collaboration, and decision making: a realist process evaluation protocol. *Implementation Science*, 9, 52.

Riek, L. D., Adams, A., & Robinson, P. (2011, March). Exposure to cinematic depictions of robots and attitudes towards them. In *W: Proceedings of the ACM/IEEE International Conference on Human-Robot Interaction. Workshop on Expectations and Intuitive Human-Robot Interaction*, March 6–9, 2011, Lausanne, Switzerland.

Robert Jr, L. P., Alahmad, R., Esterwood, C., Kim, S., You, S., & Zhang, Q. (2020). A review of personality in human–robot interactions. *Foundations and Trends® in Information Systems*, 4(2), 107–212.

Robinette, P., Howard, A., & Wagner, A. R. (2017). Conceptualizing overtrust in robots: why do people trust a robot that previously failed? In W. Lawless, R. Mittu, D. Sofge, & S. Russell (Eds.), *Autonomy and Artificial Intelligence: A Threat or Savior?*. Springer, Cham, DOI: 10.1007/978-3-319-59719-5_6.

Rodriguez-Guerra, D., Sorrosal, G., Cabanes, I., & Calleja, C. (2021). Human-robot interaction review: challenges and solutions for modern industrial environments. *IEEE Access*, 9, 108557–108578.

Różańska-Walczuk, M., Pochwatko, G., Świdrak, J., Możaryn, J., & Kukiełka, K. (2016). Wybrane predyktory postawy wobec robotów społecznych. *Prace Naukowe Politechniki Warszawskiej. Elektronika*, 195(1), 15–24.

Sanders, T., Oleson, K. E., Billings, D. R., Chen, J. Y. C., & Hancock, P. A. (2011). A model of human–robot trust: theoretical model development. *Sage Journals*, 55(1), 1432–1436, DOI: 10.1177/1071181311551298.

Scheutz, C., Law, T., & Scheutz, M. (2021). Envirobots: how human–robot interaction can facilitate sustainable behavior. *Sustainability*, 13(21), 12283.

Schmidt, R., Möhring, M., Härting, R. Ch., Reichstein, Ch., Neumaier, P., & Jozinović, P. (2015). Industry 4.0 — potentials for creating smart products: empirical research results. In International Conference on Business Information Systems, Springer, Cham, Germany, 2015; Volume 12, pp. 16–27.

Setter, T., Gasparri, A., & Egerstedt, M. (2017). Trust in multi-agent networks: from self-centered to team-oriented. In *American Control Conference (ACC)*, 24–26 May 2017, 997–1002, DOI: 10.23919/ACC.2017.7963083.

Simon, O., Neuhofer, B., & Egger, R. (2020). Human–robot interaction: conceptualising trust in frontline teams through LEGO® Serious Play®. *Tourism Management Perspectives*, 25(100692), 1–19, DOI: 10.1016/j.tmp.2020.100692.

Solomon, M. R. (2006). Zachowania i zwyczaje konsumentów. Helios, Gliwice, 242.

Vido, M., Scur, G., Massote, A. A., & Lima, F. (2020). The impact of the collaborative robot on competitive priorities: case study of an automotive supplier. *Gestão & Produção*, 27, e5358.

Wang, X., Krumhuber, E. G., & Gratch, J. (2018). The interpersonal effects of emotions in money versus candy games. *Journal of Experimental Social Psychology*, 79, 315–327.

Wasielewska, A., & Łupkowski, P. (2021). Nieoczywiste relacje z technologią. Przegląd badań na temat ludzkich postaw wobec robotów. *Człowiek i Społeczeństwo*, 51, 165–187.

Wojciszke, B. (2011). *Psychologia społeczna*, Wydawnictwo Scholar, Warszawa.

Yanco, H. A., & Drury, J. (2004, October). Classifying human-robot interaction: an updated taxonomy. In *2004 IEEE International Conference on Systems, Man and Cybernetics (IEEE Cat. No.04CH37583)*, The Hague, Netherlands, 2004, DOI: 10.1109/ICSMC.2004.1400803.

6 Human-robot interaction – ethical and communicative determinants of trust

Urszula Widelska, Joanna M. Moczydłowska and Agnieszka Tul-Krzyszczuk

Introduction

Human-robot interaction (HRI) is a multidisciplinary field that explores the interactions between humans and robots. In particular, robotics, autonomous systems, and artificial intelligence pose a number of new communication and ethical challenges. Trust plays a crucial role in these interactions, as it directly influences how humans perceive and interact with robots. Understanding the dynamics of social interactions generated by robot-human encounters requires in-depth scientific exploration.

One of the key challenges in HRI is effective communication (Peschel and Murphy, 2012, pp. 53–62; Cominelli et al., 2021, p. 9687). As we know, interpersonal communication is based, on the one hand, on verbal communication, i.e., on conducting conversations, but in many situations, it takes place using other means of expression. These are primarily gestures, changes in body position, static facial expressions, and dynamic changes in facial expressions (Bonarini, 2020, pp. 279–285). The way a human communicates with a robot depends on the type of signal used to transmit information (electrical, mechanical, sound, optical). It is also necessary to determine the method of producing this signal (generator) and its reception. A human being uses their own senses to communicate. In interactions with other people or animals, a person naturally uses vocal communication (sound signals), which requires the sense of hearing. It should be remembered that natural language is not always used for voice communication. What is often used are other sound signals whose meaning is understood by both parties to the dialogue. Such a set of sound signals is created, for example, by an owner and their dog.

In the realm of HRI, one notable omission in the development of theories and models is the absence of an ethical framework to assess the impact and consequences of HRI research and application practices. Various ethical frameworks exist that can help identify, categorize, and assess ethical challenges. These frameworks typically involve key elements such as actors (those responsible for actions with ethical implications), actions themselves, and stakeholders (those

DOI: 10.4324/9781003586937-8

impacted by said actions). Technology has become such an important social actor that the characteristics of relationships resemble those of human relationships (Etemad-Sajadi et al., 2022, pp. 1103–1115). And, as is well known, trust is one of the main elements of any social relationship (Ejdys, 2018, pp. 980–997). Ethical issues in the development of HRI should have an interdisciplinary and systematic approach. Appropriate ethical treatment, vis-à-vis informed consent, data and privacy use, manipulation, and fraud when interacting with social robots are considered (Rosén et al., 2021). In recent years, some concerns have been raised in society about the impact of robotics. It has been noted that technology can be unpredictable and, therefore, potentially dangerous. Considering ethical issues, the following variables were proven to have the greatest impact on the intention to use robots: social cues, trust, and safety (Etemad-Sajadi et al., 2022, pp. 1103–1115). Safety issues, in this case, relate to physical, psychological, and emotional damage (Lin et al., 2014).

The aim of this chapter is to systematize knowledge on HRI in the work environment with a particular focus and the identification of research gaps in the area of communication and ethics. This chapter uses the method of a narrative review of recent literature. The following research questions were formulated:

- What are the key challenges in human-robot communication?
- What ethical challenges determine HRI?

The results obtained allow the identification of further research directions for HRI.

Human-robot communication process

In the case of robots, work on attempts to generate emotions expressed by changing "facial expressions" is very advanced (Islam et al., 2019, pp. 1581–1618). Most often, sets of artificial markers are used for this purpose, with the help of which, by changing their geometry, synthetic images with symbolic meaning are generated. Therefore, the quality of the message is somewhat limited and may cause consternation, fear, or signs of disappointment (e.g., in the case of a robot communicating with small children). In such a situation, the problem of robot-human communication becomes particularly important. For the elderly and sick, as well as children isolated due to an illness, and also for autistic children, a properly constructed robot can effectively change the conditions of their functioning and help break down existing barriers to communication with the surroundings. In this situation, interpersonal communication patterns may be a good starting point for work on improving human-robot communication.

Psychologists believe that most interactions in interpersonal communication are carried out non-verbally (Leathers, 2009). For this reason, robot designers try to implement both verbal and non-verbal communication mechanisms into humanoid robots. Such robots are equipped with the ability to interact with each other and implement social behaviours. Moreover, they should have mechanisms for understanding human behaviour and some kind of empathy that allows them to express their behaviour in an appropriate way. Currently, there is great interest in the so-called personal robots. Such robots can be used in everyday life to perform heavy work or to help elderly and disabled people. Both applications require robots capable of communicating with people in a natural way, without limiting only to verbal and written communication, for example, instructional communication (Hambuchen et al., 2021, pp. 265–272). It should be emphasized that additional aspects such as gestures, facial expressions, and movement also play a very important role. In terms of construction, meeting these expectations is difficult because it is not easy to make a lightweight structure, for example, a life-size eye mechanism with cameras inside the eyeballs with comparable movement capabilities and appropriate dynamics as in humans (Zhang et al., 2021, pp. 13319–13327). Integrating both special artificial sensory systems and facial expression creation components is a very complex problem (Vasconez et al., 2019, pp. 35–48).

Another important feature of the human-robot communication process is the correct interpretation of human emotional states. It is necessary to identify movements of the trunk, neck, eyes, and especially the face (Luo et al., 2019). Correct recognition and interpretation of such movements using sensory systems is an important research and design problem. In addition to the design of the robot's mechanical systems, which should ensure clear human-robot and robot-human communication and an appropriate sensory system for collecting information about the environment, the robot needs an efficient and effective control system that ensures dynamics adequate to the conditions of the task being performed (Randell et al., 2014, p. 52; Levine et al., 2018, pp. 421–436). Work in this area is very advanced (Raj and Kos, 2022). Due to the fact that emotions are integral components of human interactions, significant efforts have been made to develop emotionally responsive robots. An example is a robot called Ameca, which is able to imitate human facial expressions in a very realistic way. Ameca can smile, frown, blink, and open and close its mouth. It may also show surprise, frustration, or amusement (Bogue, 2022, pp. 667–671). It should be clearly emphasized here that there is still an asymmetry between the ability of robots to generate spoken language and expressive behaviour compared to their ability to capture the situational context, follow, and participate in natural dialogue or read people's intentions (Lee et al., 2020, pp. 612–616). Their capabilities in this area are in contrast to the trajectory of human development in which social understanding usually precedes speech. A social robot can talk like an adult and yet have less situational understanding than a two-year-old. This imbalance, which can be confusing for users, will decrease as the robot's social sensitivity to context improves. Meanwhile, verbal HRIs are most effective in limited conditions, with clearly defined interaction goals. Reducing the production capacity of social robots to better match their cognitive advancement may also be an effective strategy. For example, animaloid robots can serve in therapeutic roles, while most are non-verbal and have limited situational awareness. An animal-like appearance may also generate significantly lower user expectations compared to humanoid robots (Prescott and Robillard, 2021, pp. 1–5).

People can spontaneously form social-emotional bonds with robots, even those that are not specifically designed to induce social behaviour (Hoffman, 2019, pp. 209–218). This is proved by the cases of emotional attachment among owners of house-cleaning robots and soldiers who cooperate with robots to dispose of bombs (Carperter, 2015). An extremely significant event was the ceremony that took place in 2015 in a Buddhist temple in Japan. It was a ceremony of dismantling a robot dog, analogous to those which priests in Japan organize for deceased pets (Prescott and Robillard, 2021, p. 4). These are just examples illustrating the potential complexity of the human-robot relationship and the ethical dilemmas that arise in this context.

Ethical aspects of the human-robot relationship

It is reasonable to ask whether there is a need to address ethical issues arising from the interaction of robots with humans. Does dealing with these problems automatically mean the emergence of a partially autonomous subdiscipline of ethical considerations, which is roboethics? In other words, do we need roboethics today? In the authors' opinion, the answer to this question is yes. This is mainly due to the complexity and diversity of HRIs. The main areas of interest of roboethics representatives include

- ethical systems built into robots,
- ethical systems used by robots,
- ethical systems used by engineers when constructing robots,
- ethical systems used by robot users,
- ethical systems used by robot manufacturers,
- the ways in which people treat robots (Coeckelbergh, 2022).

Human-robot cooperation is visible not only in the areas of production, cleaning, and entertainment. HRI also occurs with particularly vulnerable populations (Bryant et al., 2020, pp. 13–21). For example, robots have been used for therapeutic and/or diagnostic purposes for many years. Extensive scientific literature illustrates the use of robots in the treatment of children exhibiting autism spectrum behaviours (Feil-Seifer and Matarić, 2011, pp. 24–31). Robots are effectively used to reduce social isolation and loneliness in people who are hospitalized or in care facilities (Gulrez et al., 2016, pp. 1–2). The positive impact of robots in the process of therapy of people with mental disorders, especially in the field of emotional support, has also been confirmed (Kabacińska et al., 2021, pp. 919–935; Robinson et al., 2013, pp. 661–667). The effectiveness of robots as social companions is visible also in the process of health and social rehabilitation of people with dementia (Perugia et al., 2020, pp. 926–943). Social robots can be used to support social behaviour, both between the user and the robot, and by acting as a catalyst for human-human interaction. This function was confirmed by the research results of A.K. Ostrowski et al. (2019, pp. 59–70). They proved that robots can act as companions to older people, but also as robots initiating conversations between nursing home residents and drawing them into the community space. Moreover, they can improve the quality of life of older people by positive influence on their activity and regularity of taking medications and also by minimizing anxiety and stress and reducing the feeling of loneliness (Pu et al., 2019, pp. e37–e51). Robotic assistance with everyday activities, such as bathing, may also be important.

The described aspects of the HRI raise a number of ethical threats. First of all, attention is drawn to the high risk of emotional involvement of people who begin to have the same feelings for the robot as for the other people, including those closest to them. We must not lose sight of the fact that robots are designed as machines, and therefore, it is at least risky, if not completely wrong, to encourage people to treat them in the same way as social entities (Gu et al., 2019). Only other living beings (mainly humans and some animals) are able to establish authentic emotional relationships with humans. This is why there are increasingly frequent accusations against manufacturers of social robots of "false advertising" overtly causing social and emotional reactions in people (Sharkey and Sharkey, 2021, pp. 309–316). There are arguments that creating social bonds with robots may pose a threat to establishing a moral obligation toward them, which is contrary to the interests of human well-being. The robot's emotions are, in fact, a kind of "counterfeit" of real emotional bonds (van Wynsberghe and Comes, 2020, pp. 43–53). Meanwhile, people who, due to their physical and/or mental condition, engage in a relationship with a robot, for example, children or people with mental disorders, may experience strong negative effects of such a situation, e.g., a feeling of being deeply hurt or being deceived. Therefore, feeding false expectations regarding the emotional context of interaction with a robot among representatives of sensitive groups should be considered definitely unethical. From this perspective, the most pressing ethical questions concern the balance of benefits and harms that may result from allowing robots, which humans are ready to recognize as social, into our lives (Clarke, 2014, pp. 286–305). The list of potential threats and harms continues and particularly concerns human dignity, the possibility of reducing or losing contact with other people as a result of replacing this contact with interactions with a robot, and the broader emotional social robot effects (van Wynsberghe, 2015). Particular emphasis is placed on the fact that interactions with robots are basically easier and do not require empathy and commitment, so there is a risk that they will gradually replace real, complex interpersonal relationships (Bryson, 2010, pp. 196–200). The ethical dimension of the analysed interaction also concerns the involvement of robots in particularly intimate activities, such as bathing and using sanitary facilities. Even direct physical contact between robots and humans, such as when lifting a patient from a bed and into a wheelchair, creates a high likelihood of creating an emotional bond, especially in environments where human companionship is lacking. Therefore, the use of robots that physically assist people must take into account their privacy, for example, by deactivating video monitors during intimate procedures.

The answer to ethical challenges is codes that are intended to eliminate the presented risks and threats. The most famous and, at the same time, one of the first proposals regarding robot ethics was developed by I. Asimov and went down in history as the Three Laws of Robotics. They read as follows:

1 A robot may not injure a human being or, through inaction, allow a human being to come to harm.
2 A robot must obey orders given to it by human beings except where such orders would conflict with the First Law.
3 A robot must protect its own existence as long as such protection does not conflict with the First or Second Law.

In his later publications, Asimov expanded the list of principles and added a new, even more, fundamental law, which he called "Zeroth Law". It is to be superior to the other three. Zeroth Law states that a robot may not harm humanity or, by inaction, allow humanity to come to harm (Moczydłowska, 2023).

There are robot ethics codes alternative to Asimov's. One of them was developed by an international team of scientists belonging to EURON (European Robotics Research Network). This code resolves ethical dilemmas on two levels: developing ethical principles for scientists constructing robots and "artificial" ethics that will be programmed into robots. The EURON code does not yet deal with the formulation of specific laws or principles, focusing on general recommendations that can then be used as the basis for more detailed regulations. The following priorities have been identified:

- manufacturers must build machines in such a way that they are always under human control; illegal use of machines cannot be allowed;
- data obtained by robots must be protected;
- an understandable system for machine identification should be developed, as well as a method for machine tracking (Sharkey, 2014, pp. 63–75).

In other words, these are norms intended to prevent possible immoral or illegal use of machines, or to respond effectively if they suddenly "get out" of human control (Li and Meng, 2022). As R. Campa vividly writes,

> The robot is not afraid of arrest or punishment, it would have much fewer psychological and operational scruples than a human murderer, and therefore, it would definitely be more dangerous. As costs came down, more and

> more criminals would start entrusting the execution of their criminal plans to machines. This is why it would be necessary to inventory and catalogue all robots in circulation and, at the same time, shift protection and liability for their actions onto their owners, as well as establish sanctions in the event of selling or purchasing a robot not registered in the database.
>
> (Campa, 2011, p. 89)

Notwithstanding the principle of harmlessness, robotics research should be conducted in accordance with the highest standards of ethics and professionalism and ensure compliance with an additional three principles:

- bring benefits – robots should serve the best interests of humans,
- autonomy – the ability to make a conscious, unforced decision about the rules of interaction with robots,
- equity – the fair distribution of benefits associated with robotics, and in particular, the affordability of home care and health care robots.

All the indications are that legislation needs to be developed to deal with the use of robots or artificial intelligence and to prepare for a world full of ethical, social, and legal issues related to communication, privacy, and the use of massive amounts of data (Lee and Qiufan, 2021, p. 240).

Conclusion

Effective communication and ethical issues are key challenges in HRI. The implementation of interpersonal communication mechanisms, both verbal and non-verbal, in human-robot communication, becomes particularly relevant and represents a major research challenge. Current sets of artificial indicators, which allow the generation of synthetic images with symbolic meaning, may limit the quality of the message and cause anxiety, fear, or signs of disappointment. The issue of communicating naturally with humans and correctly interpreting human emotional states by robots is also crucial. The correct recognition and interpretation of such movements by sensory systems is an important research problem. Robots that can adapt their communication style to individual preferences and context can build stronger relationships and trust with users. In addition to the mechanical design, which should ensure clear human-robot communication, it is important to provide an effective control system that provides dynamics appropriate to the task conditions. Given that emotions are an integral part of human interactions that can foster greater bonding and trust, considerable efforts are being made to develop emotionally responsive robots (e.g., humanoid robots). Work in this area is well advanced. However, there is still an asymmetry between the robots' ability to generate speech and expressive behaviour and their ability to perceive situational context, participate in natural dialogue or understand human intentions.

There is a need to address ethical issues in HRI and the emergence of roboethics as a sub-discipline. Given the complexity of HRI and the diverse applications of robots, especially among vulnerable populations, ethical concerns related to emotional involvement, privacy, and human dignity become highly relevant. Furthermore, robots need to perform reliably and predictably to gain human trust. It is essential to conduct robotics research in an ethical manner and to ensure that robots serve the interests of humans while respecting the principles of autonomy and equality. The need to conduct robotics research according to the highest standards of ethics and professionalism and to ensure compliance with the additional three principles of benefit, autonomy, and equity is indicated.

There is a need to balance the benefits and harms of introducing robots into our lives. There is a risk of establishing moral obligations toward robots, which is contrary to the interests of human welfare. There are also potential threats to human dignity and the possibility of reduced contact with other humans as a result of being replaced by interactions with a robot. This could lead to the gradual replacement of real, complex human relationships. By addressing ethical issues and optimizing communication strategies, the aim is to create harmonious and productive relationships between humans and robots.

References

Bogue, R. (2022). The role of robots in entertainment. *Industrial Robot*, *49*(4), 667–671. https://doi.org/10.1108/IR-02-2022-0054

Bonarini, A. (2020). Communication in human-robot interaction. *Current Robotics Reports*, *1*, 279–285.

Bryant, D. A., Borenstein, J., & Howard, A. (2020, March). Why should we gender? The effect of robot gendering and occupational stereotypes on human trust and perceived competency. In *Proceedings of the 2020 ACM/IEEE International Conference on Human-Robot Interaction* (pp. 13–21). Association for Computing Machinery. https://doi.org/10.1145/3319502.3374778

Bryson, J. J. (2010). Why robot nannies probably won't do much psychological damage. *Interaction Studies*, *11*(2), 196–200.

Campa, R. (2011). Kodeksy etyczne robotów: zagadnienie kontroli sprawowanej przez człowieka. *Pomiary Automatyka Robotyka*, *15*(3), 66–70.

Carpenter, J. (2015). *Culture and human-robot interaction in militarized spaces: A war story*. Taylor & Francis, Nowy Jork.

Clarke, R. (2014). The regulation of civilian drones' impacts on behavioural privacy. *Computer Law & Security Review*, *30*(3), 286–305.

Coeckelbergh, M. (2022). *Robot ethics*. MIT Press, Cambridge, MA.

Cominelli, L., Feri, F., Garofalo, R., Giannetti, C., Meléndez-Jiménez, M. A., Greco, A., & Kirchkamp, O. (2021). Promises and trust in human–robot interaction. *Scientific Reports*, *11*(1), 9687.

Ejdys, J. (2018). Building technology trust in ICT application at a university. *International Journal of Emerging Markets*, *13*(5), 980–997.

Etemad-Sajadi, R., Soussan, A., & Schöpfer, T. (2022). How ethical issues raised by human–robot interaction can impact the intention to use the robot? *International Journal of Social Robotics, 14*, 1103–1115. https://doi.org/10.1007/s12369-021-00857-8

Feil-Seifer, D., & Matarić, M. J. (2011). Socially assistive robotics. *IEEE Robotics & Automation Magazine, 18*(1), 24–31.

Gu, M., Zhao, H., & Dong, T. (eds.). (2019). *Technology and application of service robot*, Southwest Jiao Tong University Press, Chengdu.

Gulrez, T., Nefti-Meziani, S., McEvoy, P., & Hodgson, A. (2016). Loneliness kills: Can autonomous systems and robotics assist in providing solutions?. *International Journal of Swarm Intelligence and Evolutionary Computation, 5*(1), 1–2.

Hambuchen, K., Marquez, J., & Fong, T. (2021). A review of NASA human-robot interaction in space. *Current Robotics Reports, 2*(3), 265–272.

Hoffman, G. (2019). Evaluating fluency in human–robot collaboration. *IEEE Transactions on Human-Machine Systems, 49*(3), 209–218.

Islam, M. J., Hong, J., & Sattar, J. (2019). Person-following by autonomous robots: A categorical overview. *The International Journal of Robotics Research, 38*(14), 1581–1618.

Kabacińska, K., Prescott, T. J., & Robillard, J. M. (2021). Socially assistive robots as mental health interventions for children: A scoping review. *International Journal of Social Robotics, 13*, 919–935.

Leathers, D. G. (2009). *Komunikacja niewerbalna: zasady i zastosowania*. Wydawnictwo Naukowe PWN, Warsaw.

Lee, C. Y., Lee, H., Hwang, I., & Zhang, B. T. (2020, June). Visual perception framework for an intelligent mobile robot. In *2020 17th International Conference on Ubiquitous Robots (UR)* (pp. 612–616). IEEE (Institute of Electrical and Electronics Engineers).

Lee, K.-F., & Qiufan, Ch. (2021). *AI 2041: Ten visions for our future*. Edbury Publishing, London.

Levine, S., Pastor, P., Krizhevsky, A., Ibarz, J., & Quillen, D. (2018). Learning hand-eye coordination for robotic grasping with deep learning and large-scale data collection. *The International Journal of Robotics Research, 37*(4-5), 421–436.

Li, K., & Meng, Z. (2022). Ethics of robotics applications. In B. J. Jansen, Q. Zhou, & J. Ye (Eds.), *International Conference on Cognitive Based Information Processing and Applications (CIPA 2022)* (pp. 325–330). Springer.

Lin, P., Abney, K., & Bekey, G. A. (2014). *Robot ethics: The ethical and social implications of robotics*. The MIT Press, Cambridge, MA.

Luo, X., Amighetti, A., & Zhang, D. (2019, July). A human-robot interaction for a Mecanum wheeled mobile robot with real-time 3D two-hand gesture recognition. *Journal of Physics: Conference Series, 1267*(1), 012056.

Moczydłowska, J. M. (2023). *Przemysł 4.0. Ludzie i technologie*, Difin, Warszawa.

Ostrowski, A. K., DiPaola, D., Partridge, E., Park, H. W., & Breazeal, C. (2019). Older adults living with social robots: Promoting social connectedness in long-term communities. *IEEE Robotics & Automation Magazine, 26*(2), 59–70.

Perugia, G., Díaz-Boladeras, M., Catala-Mallofré, A., Barakova, E. I., & Rauterberg, M. (2020). ENGAGE-DEM: A model of engagement of people with dementia. *IEEE Transactions on Affective Computing, 13*(2), 926–943.

Peschel, J. M., & Murphy, R. R. (2012). On the human–machine interaction of unmanned aerial system mission specialists. *IEEE Transactions on Human-Machine Systems, 43*(1), 53–62.

Prescott, T. J., & Robillard, J. M. (2021). Are friends electric? The benefits and risks of human-robot relationships. *Science*, *24*(1), 1–5.

Pu, L., Moyle, W., Jones, C., & Todorovic, M. (2019). The effectiveness of social robots for older adults: A systematic review and meta-analysis of randomized controlled studies. *The Gerontologist*, *59*(1), e37–e51.

Raj, R., & Kos, A. (2022). A comprehensive study of mobile robot: History, developments, applications, and future research perspectives. *Applied Sciences*, *12*(14), 6951.

Randell, R., Greenhalgh, J., Hindmarsh, J., Dowding, D., Jayne, D., Pearman, A., ... & Kotze, A. (2014). Integration of robotic surgery into routine practice and impacts on communication, collaboration, and decision making: a realist process evaluation protocol. *Implementation Science*, *9*, 52.

Robinson, H., MacDonald, B., Kerse, N., & Broadbent, E. (2013). The psychosocial effects of a companion robot: A randomized controlled trial. *Journal of the American Medical Directors Association*, *14*(9), 661–667.

Rosén, J., Lindblom, J., Billing, E., & Lamb, M. (2021). Ethical challenges in the human-robot interaction field. *ACM Digital Library*. https://www.diva-portal.org/smash/get/diva2:1539777/FULLTEXT01.pdf (Accessed: 28.02.2024).

Sharkey, A. (2014). Robots and human dignity: A consideration of the effects of robot care on the dignity of older people. *Ethics and Information Technology*, *16*, 63–75.

Sharkey, A., & Sharkey, N. (2021). We need to talk about deception in social robotics!. *Ethics and Information Technology*, *23*, 309–316.

Van Wynsberghe, A. (2015). *Healthcare robots: Ethics, design and implementation*. Routledge, Abingdon-on-Thames.

Van Wynsberghe, A., & Comes, T. (2020). Drones in humanitarian contexts, robot ethics, and the human–robot interaction. *Ethics and Information Technology*, *22*, 43–53.

Vasconez, J. P., Kantor, G. A., & Cheein, F. A. A. (2019). Human–robot interaction in agriculture: A survey and current challenges. *Biosystems Engineering*, *179*, 35–48.

Zhang, F. Z., Campbell, D., & Gould, S. (2021). Spatially conditioned graphs for detecting human-object interactions. In *Proceedings of the IEEE/CVF International Conference on Computer Vision* (pp. 13319–13327). Virtual Event, Montreal.

7 Trust of journalists in collaborating with AI

Ethical background of working with generative artificial intelligence

Anna Musialik-Chmiel, Jerzy Gołuchowski and Agata Berdowska

Introduction

For years, journalism as indicated by many media researchers has been in crisis (Butler Breese, 2016). It is demonstrated not only in decreasing the number of recipients (readers, listeners, viewers) of media products supplied by professional journalists who move to cyberspace, in particular social media. It is also demonstrated by the poorer and poorer quality of journalists' content, the effect of which is a deeper and deeper crisis of trust in journalists. Almost two decades ago, in 2007, it was distinctly described by British journalist, reporter and publisher Ian Jack, famous for his independence and conviction about the fundamental meaning of respect as the value of deciding about trust in media (Jack, 2007). Paraphrasing the author's words, one can say that media moved from "hacks" to "hyenas", and news broadcasters are now a part of the "media industry" and sub-category of show business.

The journalism crisis has also been strengthened by social and technological transformations. The media industry has recently undergone significant changes. The emergence of many internet technologies, social media and artificial intelligence (AI) changed the way in which news is created and reported. The development of ICT technology was a challenge for the media industry and news production, but companies, to adjust to this challenge, changed their business models and developed a lot of presentation forms. Large media corporations immediately noticed the potential and necessity to use new information technologies in the development of their information services which prompted them to invest in their application. At present smaller local media are also at the stage where they have to be more efficient in their news production (Tessem et al., 2024). The research demonstrates higher and higher quality of citizen media, the number of blog readers and YT channels significantly increases (Musialik-Chmiel et al., 2025).

Undoubtedly, the development of generative AI (GAI) brings new challenges for journalism. AI can automatically generate content and create media products,

DOI: 10.4324/9781003586937-9

which until recently could be prepared only by professional journalism. It can also be a support, a virtual assistant, a partner in a dialogue, leading to joint preparation of media coverage. AI used for the automatic generation of content is undoubtedly a useful tool to prepare relatively simple, template journalistic tasks, to search for specific information, to edit, etc. Similarly, virtual assistance is useful in the preparation of media materials.

The success of AI in journalism was started in 2016 by *Heliograf* – program "author" of news published in the *Washington Post* used to write news about the summer Olympic Games in Rio de Janeiro. The journalist-bot also published tweets and posts on other social media of the journal. In this way, it provided the editors with hundreds of thousands of views and itself with continuation of collaboration writing news about presidential elections in the USA, which were held the same year.

In the situation of systematic decrease of trust, as well as higher and higher competition with social media, the broadcasters and media outlets search for new opportunities. They reach for new methods and tools, including social media and AI. A justified question arises whether AI will deepen the crisis or help overcome it. A spectacular example of using AI in *Washington Post* journal, which is an icon of quality journalism, disposed the authors of this chapter to ask journalists about trust and media coverage created by them using AI. Justified questions are: can AI replace a journalist, e.g. in looking for topics, editing columns or preparing reports in investigative journalism; can AI be a competitor for a journalist profession and is it a useful tool giving an opportunity to improve the quality of journalism and restoring high quality of the profession and what is the trust of professional journalists in Poland in AI.

The main purpose of this chapter is presentation and discussion of results of studies on trust of selected Polish journalists in using IT and collaborating with AI. The research focused on the recognition of the ethical background of working with GAI. This chapter contains the review of the literature in this regard. Next, the methodology and obtained results were described. Finally, the conclusions from the conducted qualitative and quantitative results are presented.

Review of literature

The empirical studies show the relation of AI and journalism. ICT technologies are used in journalism to search for information, collect and analyze data, verify and automatically generate contents and visual elements (Tessem et al., 2024). Although previously various forms of AI were available and applied for many years to create and distribute news and information online, at present the application of new generative forms in journalism provides significantly more possibilities, for example, to prepare reports about earnings or summaries of sports events or generate tags or for transcription (Artificial Intelligence in Journalism, 2024). GAI is used to create automated news, using highly structured and easily accessible all over the world data, results of intelligent data analysis in media and knowledge graphs, used to edit news. The news created in this way can also be published in social media.

The studies indicate numerous benefits of using GAI by editors. They include faster identification of the latest news, inspiring journalists to ask less obvious questions and automation of routine tasks such as transcript of interviews or translation of documents (Schantin, 2024). GAI can also analyze huge numbers of data to reveal stories which could have been skipped by people-journalists and personalize information channels to make sure the readers receive adequate content in the format they prefer. However, the use of AI is not free of traps. One of the main problems is algorithmic bias built-in AI tools. AI systems are as good as the data on which they are trained. If the data are biased, the AI results will also be biased, which potentially leads to misrepresented reporting (Schantin, 2024).

Understandably, the use of AI in journalism encountered both favorable interest and concerns. Until recently the concerns regarding development of AI focused on its potential abilities to replace people in physical work, however, at present, these considerations move to issues of destabilization of workplaces of "white collars", including media ("purple collars"). It is also noticeable that copyrights are misused by AI authors and suppliers, which often destabilizes the work of traditional editorial offices (regional or local) by BiGTech as well as media corporations by providing the public with mis- and disinformation, which even more undermines the trust in media (Artificial Intelligence in Journalism, 2024; Arguedas & Simon, 2023).

In the literature it is emphasized that to create and keep trust media organizations should be transparent about using AI, explaining how these systems work, which data they use and what steps they take to mitigate prejudice (Arguedas & Simon, 2023). Some organizations label the headlines of articles as generated by GAI (Schantin, 2024). In the literature, it is sometimes indicated that the key factor for AI success both in journalism and in media is finding a way to increase the trust in contents without labeling each element with information "created using AI". Legal regulations are heading in this direction. European AI Act emphasizes certain principles of informing users about AI use. Others claim that also human supervision over AI operations is important and it should be a basic part of the code of conduct of every information organization with reference to AI and GAI. In practice, it means that every editorial office should treat contents created by a machine as unverified source of information and the ultimate responsibility invariably remains on people-journalists (Report: Concerns…, 2024).

Undoubtedly trust is the foundation on which every successful media organization is built. Readers believe that the information provided by media is accurate, honest and unbiased. However, with the occurrence of AI, as already noticed, there are increasing concerns whether these technologies can keep trust standards that their recipients expect. According to the research of Edelman Trust Barometer (2024 Edelman Trust Barometer, 2024) and as described (Musialik-Chmiel et al., 2023), trust in media companies and journalism has been decreasing for a few years. Turning this trend is a key challenge for the media industry. By analyzing the research presented in the literature a conclusion can be drawn that the matter of journalists' trust in AI has never been as important as now. Due to the fact that editors more and more often use the assistance of GAI, keeping and building the trust in the face of quick technological changes is of primary significance (Schantin, 2024).

The latest reports indicate many factors shaping trust in AI. One of them is the lack of transparency how algorithms work. It raises a question about bias and reliability of such tools. The significance of this issue is, among others, indicated by the latest research of Reuters Institute (Digital News Report 2024, 2024) conducted in 47 countries of the world, which indicated that the percentage of all respondents who said they were concerned what is true and false on the Internet increased by 3% year to year from 56% to 59%. The report of Reuters Institute (Digital News Report 2024, 2024) also contains information on how, in the previous year, media companies implemented AI solutions with a various degrees of supervision from a man. For example, editors from Nordic countries place bullet points generated by AI in headlines of many articles in their publications to increase the number of readers. It was also noticed that one of German editors used AI to write over 5% of its published articles. Other publishers implemented such tools as Midjourney or Dall-E for automation of graphics. The report also demonstrated that websites (information services?) from Indonesia, South Korea, Slovakia, Taiwan and Mexico when presenting news used experimental chatbots and avatars. In addition, subjects more and more often using AI to rewrite news without permission and human supervision were identified.

The concerns related to the use of GAI in media industry also involve not respecting copyrights and potential errors caused by so-called hallucinations. This phenomenon is well documented but little is known how recipients feel these questions and how it influences the trust in authors-people. The study results indicate that recipients want to know if they are dealing with AI or a human being. Also they feel less discomfort when GAI is used to support human journalists, for example for transcription of interviews or preparing reports from research, especially regarding sensitive topics (politics, war or crime) (Digital News Report 2024, 2024). Other studies on trust in media conducted by Hofeditz et al. (2021) suggest that there is no significant difference in the level of trust in a situation when participants were previously informed about creating articles using GAI. Similarly, the reliability, readability and journalistic knowledge were assessed as irrespective of the information about GAI application. According to the results of this study, when assessing the reliability of an article, the actual content was much more important than the authorship. Moreover, social theories, such as access control theory (Nahon, 2008) and nest access control theory

(Deluliis, 2015), can have more influence on the assessment of reliability of information articles created by AI or a human journalist.

Regarding the trust in application of GAI in journalism, studies aiming at establishing whether editorial offices in their internal regulations take into consideration issues connected with their use were identified (Merrefield, 2023; Becker et al., 2023). Among elaborations there were analyses presenting results: over 71% documents mention one or more journalistic values, such as public service, objectivity, autonomy, directness (speed of publishing or broadcasting news and ethics); almost 70% documents concerning AI were intended for editorial offices but most of them referred to the entire organization, particularly in business aspects – it was admissible to use AI for advertising or recruitment purposes; 69% regulations mentioned traps of AI, such as "hallucinations" or making up facts; about 54% of such documents warned journalist about a necessity to protect sources while using AI, a few of them referred to a potential risk of revealing confidential sources while providing a chatbot with information; 44% regulations allowed journalists to use AI to gather information and work on ideas, points of view and history outlines; 42% allowed a possibility to use AI to alter editorial contents, e.g. edit and update articles. What is worth noticing – the research demonstrated that only 8% regulations specified how the rules concerning use of AI will be enforced whereas in others the mechanisms of responsibility were not mentioned at all (Merrefield, 2023; Becker et al., 2023). This question is also particularly evident both in academic and public discourse in the scope of creating AI rules, which will be trustworthy. Attempts are taken to prepare guidelines for creation of responsible an ethical AI. Unfortunately, at present these rules are too technocentric while in reality the trust is human judgment or attitude, which can be formally defined as a judgment of reliability in situations characterized by susceptibility to influences. The same AI technology can be judged in a different way by various people, and some of them formulate inprecise judgments of trust. Finally, the psychological perspective determines the way in which people will use AI and interact with it. It also decides whether they can be harmed by inappropriate trust (credulity), meaning excessive reliance on it and consequently wrong use. For authors of these guidelines (ethical AI systems), two questions are most important, which have particular meaning in holding AI technology and its creators accountable. First, malignant manipulation of guidelines concerning reliability, which may lead to undeserved trust, threatens with serious consequences. In this case it is necessary to determine the reliability of the base model and communication about it as the foundation to consider managing AI system project. The other important matter is that even creators of AI have good intentions which can accompany wrongly designed rules concerning AI reliability, which will harm users (Liao & Sundar, 2022). An example of research conducted on guidelines for creation of AI ethical systems is an elaboration prepared by researchers of Michigan State University (Choung et al., 2022). This work mentions that designing AI based on values which make it trustworthy should be a critical project problem.

Methodology

Qualitative and quantitative research conducted from 2022 to 2024 (Musialik-Chmiel et al., 2023; Musialik-Chmiel et al., 2024) was carried out in a group of Polish professional journalists who were representatives of various editorial offices (Polish national and regional media), cooperating with a lot of media present in social media, some of them are popular podcasters and YouTubers. The purpose of the research was to obtain knowledge about trust in journalists and media messages created by them in various contexts. For this purpose, questionnaires were conducted in a group of 12 journalists and 15 individual in-depth interviews. It was important to find out (1) if Polish journalists use ICT tools to verify information and precisely which tools they use in their daily work and (2) what their attitude to the application of GAI is.

Results

A variety of topics of texts which a journalist prepares requires them to acquire both knowledge of a lot of topics and proofs that what they want to say is true. That is why it has become so important to be able to find information of the internet or to contact sources of information via new technologies. It is also worth noting that despite certain positive aspects of multifunctionality such as a bigger field for creativity, the multitasking decreases the value of produced materials. From the respondents' answers emerges a desirable/plausible image of qualitative journalism, characterized by aspiration to truth and adequacy, honesty, respect for other's privacy, respect for the law, respecting good manners and good taste, and independence of influences from various groups of interests.

The research we conducted is very important in the context of application of GAI. It allows drawing initial conclusions concerning journalists' readiness to cooperate with new technology and combat negative examples of use of this type of tools. Prejudice and concerns about losing a job or unjustified belief that AI inevitably lead astray are definitely rejected by interlocutors. They claim that AI errors so far were effects of human negligence and errors. ChatGPT itself asked by a journalist if it intends to take his job replies that *Artificial intelligence does not have the possibility to create valuable journalistic contents independently. Neither does it have the ability to take moral decisions, which are often necessary in journalistic work* (IDI, 2022–2024).

The answers provided by questioned respondents allow the initial conclusion that the most popular tool of facts verification for journalists is still Google browser. Journalists, IDI participants, asked about fact-checking admitted that they rather do not actively participate in fake news verification, conducted by specialized institutions though they do not deny the need for existence of fact-checking agencies in Poland. However, it must be added that among IDI participants there were journalists involved in #fakehutner project started by PAP, dedicated, first of all, to all those who would like to check reliability of contents coming from social media.

GAI assistance in this area can bring very rational positive effects. It is also justified to assume that some journalists are not ready to cooperate with GAI like chatGPT, because they believe it is not beneficial for journalists practicing quality journalism. People invited to IDI research are connected, among others, with the magazine Gość Niedzielny (The Sunday Guest), which has a well-established position not only in the region (Upper Silesia), where it has existed for over 100 years, but it also enjoys an opinion of one of the best opinion-forming weekly magazines in Poland. The representatives of this magazine give provocative answers: *Artificial intelligence should get the Grand Press award. Not for displacing journalism, as some suggest, the opposite – it proved that quality journalism is irreplaceable. Even the most advanced version – with the bets algorithms and skills of collecting billions of data, association of styles, persons, events and authors – will never write an authentic coverage, interview, investigative text, column, original analysis or real news (IDI,* 2022–2024).

Fake news, infodemic, internet hate and disinformation are now very serious challenges for media and journalists – the more one should try to build trust in relations on many planes (professional, social, personal), so the question about using GAI in this area also seems justified, even more because it is answered by journalists-practitioners present in media for many years. One can say that they have become authorities for their recipients. They cope well with challenges on new technology, also journalists – podcasters and YouTubers. One of them, within the foundation he established "Nauka. To Lubię" ("Science. I Like It") – doctor of physical sciences and as he emphasizes himself a journalist by education - conducts workshops, lectures, writes articles and creates podcasts popularizing science, including new technologies both among adults and children. Asked about trust in GAI at is work he first of all emphasizes its utility as a tool. He sees benefits, first of all, in the scope of searching for contents, however, in his opinion it requires meticulous verification and critical analysis.

Media need ethical reflection because they affect lives of individuals and societies. Gratification for ethical attitude cannot be expected, sometimes even the price is leaving the editorial office. One person mentions a situation when conversations recorded in a radio studio were kind of automatically censored by himself – *I know that anyway it will have to be removed, because the editor will not allow its publication. It was one of very important reasons for quitting a job for the Editorial Office.* The interviews participants talk about problems which start when it comes to transmitting the truth, e.g. publish possessed materials and so to say self-censorship automatically joins the process. They all agree that the truth is a fundamental value and journalists are its guardians (Kreft et al., 2024).

Discussion

Afterthoughts of in-depth interviews participants create a kind of background for ethical aspects of cooperation with generative (conversational) AI, which is one of main topics of this chapter. They are also in line with the discussion whether technological and economic changes should be perceived as the main reasons of crises in the world of journalism, not only in Poland. In 2016 (first edition) a set of essays *The Crisis of Journalism Reconsidered*, edited by Alexander J. C., Butler Breese E. and Luengo M., was published, in which authors draw attention to the role played by cultural obligations of journalism, combining professional ethics with democratic aspirations of societies, in which journalists do their craft. Researchers ask how new technologies are created to sustain values and journalistic obligations to respect them and not challenge them. This paper offers an important contribution to a scientific debate on future of journalism, suggesting reflection on the meaning of cultural imponderables in explaining the crisis with which struggle both journalism and democracy in the era of digital media. This set of original essays analyses narration about "crisis of journalism" from a totally different perspective, emphasizing cultural codes. It is them which not only cause anxiety/concerns about technological and economic changes but also provide paths of control so that democratic practices of independent journalism can be sustained in new forms.

By emphasizing ordinariness of their professional way ("topics lie in the street", as Polish journalists say), a journalist not only informs but also explains, comments and critically analyses reality. However, he is not infallible so he inquires, asks inquisitive, uncomfortable questions, it is a moral obligation to ask them on behalf of recipients. In integral ethics, they refer to the nature of things and their identity ("what, who is it?"), to purposefulness and meaning ("what purpose something is for?"), motives and justifications of actions ("why something is?") and to the way of existence and operation ("in what way does something exist and operate?"). These questions and attempts to answer them include the need to search for truth, also in the scope of media ethics, thus also journalistic ethics and deontology. In this aspect, GAI cannot be an obstacle – "*we will always need a real reason, alive, not artificial. No artificial intelligence will make it up" – says a journalist* (IDI, 2022–2024).

The respondents' belief, also present in other studies, that GAI will not replace a journalist in conducting investigative or scientific journalism or writing a column, can be shared. The qualitative journalism will remain a remedy for erosion of this profession and this view is not common in the Polish world of journalism. Nevertheless, it is not a reason for self-satisfaction but the opposite. GAI is a challenge which everyone doing this job has to face regardless of seniority and age. Maybe over time journalists will learn in interaction to collaborate with GAI treated as a virtual intelligent assistant, after all AI is developing dynamically. GAI will not replace brilliant journalists but it can become their partner in creating more perfect texts/media products.

In journalism, even the most advanced GAI does not resolve the challenges of conducting qualitative journalism. It is exemplified by journalism in Poland – deficiencies of implementation of the mission by public media in

Poland. Vast alterations introduced in media legislation (novelization of the act on radio broadcasting and television) in July 2018 were intended to open a new chapter in functioning of public media law in Poland, the proof of which is, among others, "The card of mission duties" (2020). The implementation of such first document for the years 2020–2024 was to be completed by broadcasters in 2024. Despite the prepared regulatory activities, a gap in efficiency of the introduced changes aimed at more effective implementation of the mission by public media is noticed. Political turbulence does not promote creation of media orderliness. The opinion about bad condition of media in Poland is not isolated, just the opposite, more and more often in public discourse arises a question about the sense of public media existence and future of independent journalism.

The situation which occurred in Poland demonstrates that AI in journalism will not fill in the gap of insufficiency of values creating the media orderliness. And they should be the foundation/support of qualitative journalism, the guardian of values and incubator for journalistic talents (Freedman, 2019). Practicing journalism in conformity with normative theory of qualitative journalism, not just informing but also forming critical thinking and ethical acting is a warranty of creating professional standards and high quality of the profession as well as creation and continuation of good practices. They create principles which are fundamental in this profession: aspiration for truth, reliability, professionalism and respect for facts and recipients, and the phenomenon of the world of new technologies – AI is the reality which a journalist has to face. Can specialized AI tools guarantee implementation of values and impartiality in creation of journalistic productions?

Conclusion

The development of GAI brings new legal and ethical challenges regarding how information organizations and journalists use AI in production and distribution as well as the manner in which AI systems use information contents for learning. According to participants in the IDI, the use of GAI tools within editorial offices can enhance productivity and foster innovation motivating journalists to engage in what is commonly called "brainstorming". However, respondents expressed concerns that these tools can lead to inaccuracies and ethical dilemmas, ultimately undermining public trust.

To meet this challenge, the regulations will have to offer clear definitions of AI categories and specific exposures for each of them. They also have to face after-effects of contents generated by AI in the case of (1) violation of copyrights or terms and conditions of using the service and (2) citizen freedoms, which probably in practice will be difficult to identify and execute. Publishers and technological companies will also be responsible for establishing clear, ethical guidelines/principles and education in this scope. The future collaboration between decision-makers, publishers, technology creators and scientific workers has a key meaning (Artificial Intelligence in Journalism, 2024).

In media organizations which care for the trust of their clients, there should be mechanisms of responsibility in order to quickly and efficiently solve errors of people connected with media (media people), which will provide the readers and viewers with trust in reliability of received information, analyses and comments. It is necessary to ensure that contents generated by AI are in conformity with the same ethical standards as contents generated by people (Schantin, 2024).

The research results presented in this chapter allow determination of further deepened studies on trust in journalists' collaboration with AI. The authors intend to conduct more detailed research also in the international context. A justified question arises if, so far, AI does not use coherent axiological system then replacement of a journalist in his/her primary function is little probable. However, it cannot be excluded that sooner or later AI will be equipped with the possibility to make assessment including moral one, however, it is hard to believe that such assessment will be in conformity only with one of many axiological systems so far generated by human cultures in *Longue durée*. Until it happens, the ethical role of qualitative journalism will still be socially useful and the meaning of collaboration with AI based on trust of journalists in collaborating with AI will probably increase. The condition is development of ethical background of working with GAI.

References

2024 Edelman Trust Barometer. (2024). Edelman Trust Institute. Retrieved from: https://www.edelman.com/sites/g/files/aatuss191/files/2024-02/2024%20Edelman%20Trust%20Barometer%20Global%20Report_FINAL.pdf

Arguedas, A. R., & Simon, F. M. (2023). *Automating Democracy: Generative AI, Journalism, and the Future of Democracy*. Balliol Interdisciplinary Institute, University of Oxford, Oxford, United Kingdom, p. 21.

Artificial Intelligence in Journalism (2024, July 23). *How Do We Enable the Benefits and Manage the Harms of Artificial Intelligence in Journalism?*. Center for News Technology & Innovation. Retrieved from: https://innovating.news/article/ai-in-journalism/

Becker, K. B., Simon, F. M., & Crum, C. (2023, September 7). Policies in Parallel? A Comparative Study of Journalistic AI Policies in 52 Global News Organisations. https://doi.org/10.31235/osf.io/c4af9. Retrieved from: https://osf.io/preprints/socarxiv/c4af9

Butler Breese, E. (2016). The perpetual crisis of journalism: Cable and digital revolutions. In Alexander, J. C., Butler Breese, E., & Luengo, M. (Ed.), The Crisis of Journalism Reconsidered: Democratic Culture, Professional Codes, Digital Future. Cambridge University Press, New York, pp. 31–43.

Choung, H., David, P., & Ross, A. (2022). Trust in AI and its role in the acceptance of AI technologies. *International Journal of Human–Computer Interaction*, 1–13. https://doi.org/10.1080/10447318.2022.2050543.

Deluliis, D. (2015, March). Gatekeeping theory from social fields to social networks. *Communication Research Trends*, 34(1). The Jesuit Community at Santa Clara University, d/b/a SCJC Communication Research Trends.

Digital News Report 2024. (2024). Reuters Institute & University of Oxford. Retrieved from: https://reutersinstitute.politics.ox.ac.uk/digital-news-report/2024

Freedman, Des J. (2019). "Public service" and the journalism crisis: Is the BBC the answer? *Television & New Media*, 20(3), pp. 203–218.

Hofeditz, L., Mirbabaie, M., Holstein, J., & Stieglitz, S. (2021). Do You Trust an AI Journalist? A Credibility Analysis of News Content with AI-Authorship. ECIS 2021 Research Papers 50. Retrieved from: https://aisel.aisnet.org/ecis2021_rp/50

IDI. (2022–2024). In-depth interviews by chapter in 2022–2024. Interview transcripts available in the authors' archive.

Jack, I. (2007). *The Granta Book of Reportage*. Portobello Books, Granta Books, London, p. 27.

Karty powinności spółek mediów publicznych - obowiązujące w latach 2020-2024. (2020) Krajowa Rada Radiofonii i Telewizji. Retrieved from: https://www.gov.pl/web/krrit/karty-powinnosci-spolek-mediow-publicznych-aktualnie-obowiazujace?page=1&size=10

Kreft, J., Boguszewicz-Kreft, M., & Fydrych, M. (2024). 'Guardians of the truth' - journalists' resistance to the algorithmization of journalism. *Ethos*, 37(2 (146)), 132–149. https://doi.org/10.12887/37-2024-2-146-08

Liao, Q. V., & Sundar, S. S. (2022). Designing for responsible trust in AI systems: A communication perspective. In *2022 ACM Conference on Fairness, Accountability, and Transparency* (FAccT '22), June 21–24, 2022, Seoul, Republic of Korea. ACM, New York, NY, USA, 18 pages. https://doi.org/10.1145/3531146.3533182.

Merrefield, C. (2023, December 12). Researchers compare AI policies and guidelines at 52 news organizations around the world. *The Journalist's Resource*. Retrieved from: https://journalistsresource.org/home/generative-ai-policies-newsrooms/

Musialik-Chmiel, A., Gołuchowski, J. & Berdowska, A. (2023). Building trust in professional business journalism in the age of digital media and disinformation spreading. In Paliszkiewicz, J., & Guerrero Cusumano, J. L. (Ed.), *Communication, Leadership and Trust in Organizations*. Routledge, New York. https://doi.org/10.4324/9781003370390

Musialik-Chmiel, A., Gołuchowski, J. & Berdowska, A. (2025). Reliability of financial influencers and its effect on generation Z investors' behaviours and decisions. In Paliszkiewicz, J., Gołuchowski, J., Zdanowicz-Cyganiak, K., & Chen, K. (Ed.), *Trust, Media and the Economy Mutual Relations*. Routledge, New York. ISBN 9781032874739

Musialik-Chmiel, A., Gołuchowski, J. & Berdowska, A. (2024). Reliability of financial influencers and its effect on generation Z investors' behaviours and decisions. In Paliszkiewicz, J., Gołuchowski, J., Zdanowicz-Cyganiak, K., & Chen, K. (Ed.), *Trust, Media and the Economy Mutual Relations*. Routledge, New York.

Nahon, K. (2008, July). Toward a theory of network gatekeeping: A framework for exploring information control. *Journal of the American Society for Information Science and Technology*, 59(9), 1493–1512. https://doi.org/10.1002/asi.20857

Report: Concerns over Generative AI in News. (2024, July 17). Advanced Television. Retrieved from: https://advanced-television.com/2024/07/17/report-concerns-over-generative-ai-in-news/

Schantin, D. (2024, July 11). *AI identifies breaking news faster, managers routine tasks, but trust issues remain.* INMA Blog. Retrieved from: https://www.inma.org/blogs/media-leaders/post.cfm/ai-identifies-breaking-news-faster-manages-routine-tasks-but-trust-issues-remain

Tessem, B., Tverberg, A. & Borch, N. (2024). The future technologies of journalism. *Procedia Computer Science*, 239, 96–104. https://doi.org/10.1016/j.procs.2024.06.151

8 Psychological factors affecting swift trust in AI

A comparison of Poland and Turkey

Fatih Çetin, Anna Jasiulewicz, Selay Ilgaz Sümer and Joanna Paliszkiewicz

Introduction

Trust is commonly perceived as a psychological mechanism for mitigating uncertainty and improving the probability of successful interactions with entities in one's surroundings (Paliszkiewicz & Gołuchowski, 2025). Trust in technology is shaped by human attributes, environmental factors, and technological aspects (Siau & Wang, 2018). Given that artificial intelligence (AI) offers numerous novel features compared to other technologies, scientists started to concentrate on various possible predictors of trust in AI. Establishing trust in AI poses considerable challenges. Unlike humans, AI lacks awareness, emotions, and personal accountability, making it difficult to apply the same trust indicators utilized in human interactions. AI operates on algorithms and data, making decisions based on patterns rather than a genuine understanding of human feelings or intentions. The concern about trust in AI is becoming increasingly significant due to rising skepticism, as AI presents numerous challenges in fostering and sustaining trust (Paliszkiewicz & Gołuchowski, 2025, Makiridakis, 2017). Siau and Wang indicate that "trust is the cornerstone of humanity's relationship with artificial intelligence. Like any type of trust, trust in AI takes time to build, seconds to break, and forever to repair once it is broken".

Recent studies have revealed that trust in AI is affected by the attributes of the AI itself as well as the traits of its users (Bach et al., 2022). In this chapter, we focus on swift trust which is a kind of presumptive form of trust in a short-term period toward people working virtually or artificial agents (Robert et al., 2009). Our considerations center on exploring specific psychological factors that may influence the formation of swift trust in AI, namely, trust propensity, social norms, human-like AI perception, and AI functionality.

This chapter is structured into six sections. Following the introduction, the second section provides a brief literature review focusing on chosen key psychological factors. In the third section, we present the proposed hypotheses, which are subsequently tested later in this chapter. The fourth section details the methodology employed in our study, explaining the research process, sample

DOI: 10.4324/9781003586937-10

demographics, and questionnaire design, including the measurement instruments and the reliability of constructs. The fifth section presents the research findings, focusing on hypothesis testing and addressing the research question. Finally, the sixth section offers a discussion of the findings, along with conclusions, limitations, implications, and potential directions for further research.

Literature review

Swift trust in AI

The concept of trust has been associated with competence or ability, benevolence, and integrity considering trusting in people (Mayer et al., 1995). Interpersonal trust involves assessing whether the other party possesses the necessary qualities and intentions to act as anticipated. From a technological standpoint, the trusting factors are transforming into functionality, helpfulness, and reliability or predictability based on the system characteristics (McKnight et al., 2011). Trust in technology focuses on evaluating the characteristics of the technology itself rather than its underlying motives. However, in relatively temporary relations or in trusting an artificial agent, social structures and actions are more emphasized by establishing a swift trust. As it was mentioned in the introduction part, swift trust is a type of provisional trust established over a short-term period toward individuals working virtually or artificial agents (Robert et al., 2009). In this type of trust formation, people import social structures and actions from other contexts to influence others. Some of these constructs are dispositional-based trust, role-based trust, third-party recommendation–based trust, and category-based trust (Kramer, 1999). The base of dispositional-based trust (*propensity of trust*) is that individuals differ considerably in their general predisposition to trust other people. The role-based trust bases on the knowledge that a person occupies a particular role in the organization. It is a form of depersonalized trust that depends on the expectations of technical competence, fulfillment of responsibilities, and carrying out obligations for performing well (*functionality*). The third-party recommendation–based trust is based on disseminating trust-related information as a kind of "second-hand" knowledge about the trustee (*social influence*). Category-based trust refers to information about a trustee's membership in a social or organizational category in which the trustor is (*human-like*).

Propensity of trust

The main factors of trust are cognitive-based (functionality or reliability), affect-based (helpfulness), and context-dependent (trustor's general propensity to trust) (McAllister, 1995). General trusting beliefs typically consist of people's propensity to trust others or technology, faith in humanity, or general technology

(McKnight et al., 2011). From a nomological perspective, specific trust in persons or technology has been affected by general trusting beliefs as the predictor of trusting intentions. People's general tendency to trust influences their beliefs to trust others and develops with social learning cause of the collection of past experiences (Kramer, 1999). The collection of past experiences forms the dispositional propensity to trust that drives trusting thoughts and beliefs when there is no or limited information about the trustee (Alarcon et al., 2018). The propensity for trust denotes a personal trait, indicating one's general inclination to place trust in technology.

Functionality of AI

The similarity between a person's competence and a technology's functionality lies in their embodiment of users' expectations regarding the trustee's capacity and expertise. The inherent characteristics of the technology form individuals' trusting beliefs. The capacity or capability of a technology to fulfill a particular task defines its functional aspect (McKnight, 2005). Functionality constitutes a knowledge-based trust in which users collect information on its properties and observe its reactions to different actions by engaging with technology across various settings (Mcknight et al., 2011). However, unlike traditional technologies, AI operates with varying degrees of autonomy rather than depending solely on user input and human-programmed rules. The autonomous operation presents users with risks and uncertainties, highlighting the importance of trust as users cope with the complexities and potential risks. AI's distinctiveness lies in its combination of functionality with human-like abilities (Krafft et al., 2020).

Social influence

Social influence refers to the degree to which people are influenced by others to conform to fit with the group (Deutsch & Gerard, 1955). Based on the theory of reasoned action (Fishbein & Ajzen, 1975), social norms stem from the influence of reference groups and positively affect individual intentions and behaviors. As a kind of social normative pressure on people's beliefs, attitudes, and behaviors, social influence has been found to have a positive impact on trusting new technologies (Venkatesh et al., 2003). In addition, when limited or no knowledge in understanding of a system, people often view collective beliefs as a dependable and trustworthy source of information (Althuizen, 2018). Therefore, social influences stemming from individuals' social network advocates using AI are likely to have an impact on people's trust beliefs in AI technologies.

Human-like AI perception

The impressions based on appearance play a crucial role in fostering trust between humans and AI technologies (Todorov et al., 2015). Anthropomorphism, the act of assigning human characteristics, motivations, intentions, emotions, or behaviors to nonhuman entities (Epley et al., 2007), is increasingly pivotal in establishing trust. As users anthropomorphize nonhuman entities and embrace a self-centered view, perceiving them as "like me," the anthropomorphic perception of AI is closely intertwined with user identity and self-concept (MacInnis & Folkes, 2017). Human-like AI perception refers to the algorithms' social and cultural values and the ethics underpinning AI technology design. This perception of similarity fosters an attraction akin to social identification, resulting in increased likability, trust, and other positive attributes (Tajfel, 1978).

Hypothesis development

Propensity to trust is assumed to be linked to self-reported trust (Faulhaber et al., 2021) as well as reliance. It is regarded as a key element in the development of trust (Lee & See, 2004). According to Küper and Krämer (2024), the propensity to trust was one of the crucial factors recognized as influencing trust. Moreover, they consider that the propensity to trust technology refers to a person's belief that automation is reliable and trustworthy. Lee and See (2004) suggest that a higher propensity to trust enhances the ability to determine when to rely on or disregard automation advice. The relationship between swift trust and AI is crucial in ensuring safe and trusted AI (Ganapati & Desouza, 2025). According to Scholz et al. (2024), the propensity to trust was shown to be a strong predictor of decisions in the trust game involving both human and robotic agents. They proposed defining the propensity to trust in automated technology (PTT-A) as a more specialized personality trait that applies to different types of automated systems, such as robots, self-driving vehicles, and other AI-driven technologies. They contended that, from a psychological standpoint, the cognitive and emotional mechanisms behind the propensity to trust humans and PTT-A are essentially the same. Therefore, the following hypothesis is proposed.

H1: Propensity to trust has a positive effect on swift trust in AI

Previous studies indicate that functionality is a key dimension of trust in technology. Functionality pertains to the technology's capabilities, which the authors compared to human abilities (McKnight et al., 2011). McLean and Osei-Frimpong (2019) discovered that functional advantages (specifically, perceived usefulness and ease of use) positively affect users' engagement with AI virtual assistants. Their findings indicate that perceived usefulness and ease of use are key factors that shape the development of consumer trust. The study of Zang et al. (2021) investigated the impact of the functionality on trust through two aspects: perceived usefulness and perceived ease of use of AI virtual assistants. It confirmed that both dimensions were significantly and positively associated with trust. This might suggest that a seamless service experience fosters trust and motivate users to engage with AI more frequently, thereby cultivating trust through familiarity with AI capabilities. Hence, we propose the following hypothesis.

H2: Functionality of AI has a positive effect on swift trust in AI

Social influence is regarded as affecting the development of the internal belief system that determines an individual's behavioral intentions (Maruping et al., 2017). However, Barbul and Bojescu (2023) found out that individuals tend to place little importance on the opinions or beliefs of others regarding the use of AI. Inaba and Togawa (2021) identified four types of social capital: cognitive social capital and structural social capital—namely, contacts with others, group participation, and workplace social capital—as potential influences on the perception of AI. They found that the indirect effect appears to be primarily driven by two forms of structural social capital: social capital through group participation and in the workplace. Additionally, cognitive social capital is directly and positively associated with AI perception, whereas structural social capital, in the form of personal contacts, is negatively associated with AI perception. Individuals with higher levels of personal contact tend to be more cautious toward AI. Based on the above, we recognized the need to propose and test the following general assumption.

H3: Social influence on AI has a positive effect on swift trust in AI

The progressive evolution of computers toward enhanced empathy in human interactions—becoming more human-centered, anticipatory, and pervasive—brings to light fundamental questions about the relationship between humans and technology. In particular, it underscores the issue of the nature of trust users place in empathic technologies (Moritz, 2017). Anthropomorphic characteristics have been extensively examined in numerous previous studies across a variety of research contexts (e.g., robots and virtual social agents: Jacquet et al., 2019; Pelau et al., 2021; Liu & Tao, 2022). Scholars have varying perspectives on the relationship between AI humanization and AI acceptance and trust in AI. A lot of studies have also reported a positive influence of anthropomorphism on willingness to use advanced technologies (i.e., Roy & Naidoo, 2021; Pelau et al., 2021, Liu & Tao, 2022) For instance, Mayrashi's (2024) research found that perceived humanity of AI has a positive impact on public sector managers' trust in AI systems. However, according to some researchers, the perception of AI as human-like may impede the adoption process of AI (Tinwell et al., 2011; van Doorn et al., 2017, Kim & McGill, 2018). A moderate perception of humanity in AI virtual assistants can enhance trust between human users and the technology. Nevertheless, an overly human-like perception may cause users to feel threatened or fearful, potentially leading them to discontinue their interactions (Zang et al., 2021). Overall, these mixed findings indicate a need for further research to deepen understanding of the impact of human-like AI on trust in AI. Since previous literature has shown that AI humanization positively affects users' attitudes and intentions to use AI, we therefore anticipate a positive relationship between human-like AI and trust. Therefore, we put forward the following hypothesis.

H4: Human-like AI has a positive effect on swift trust in AI

Additionally, the following research question was formulated:

RQ: Are there any country-related differences (Poland and Turkiye) in the effects on swift trust in AI?

Method

Participants

Data were collected from 639 university students studying a university from each country (400 from Poland and 239 from Turkiye). The voluntariness and privacy of personal data were provided. The information was gathered with the utilization of the Computer-Assisted Web Interviewing method. The online questionnaire was prepared including items from literature with demographic details such as gender and grade. Gender distribution is 53% female and 47% male. The grade distribution percentages are 24.2% first grade, 25.7% second grade, 24.9% third grade, and 25.1% fourth grade.

Instruments

In this study, examined variables included propensity of trust, functionality, social influence, and human-like AI. All components were derived from pre-existing measurement scales, with certain scale items adjusted to enhance their relevance in the context of AI.

Propensity of trust: The four items have been used based on the research (Frazier et al., 2013) which tests different items in the literature and finds consistent higher loadings in four consecutive studies. An example item is "Trusting another person is not difficult for me".

Functionality: The seven items have been used based on the research (Qin et al., 2020) which proposes different technology-related factors from qualitative research on concerning AI. An example item is "AI technologies have an accuracy of diagnosing answers".

Social influence: The three items have been used based on the research (Qin et al., 2020) which uses the subjective norms for the technology adaptation. An example item is "People who influence my behavior think that I should use AI technologies".

Human-like AI: The three items have been used based on the research (Choung et al., 2022) which uses the human-like trust in AI for the technology adaptation. An example item is "AI technologies which have human-like intelligence are more honest than other digital technologies (Internet, IOT, Cloud, Big data, Blockchain, etc.)".

Swift trust: The three items have been used based on the research (Choung et al., 2022) which uses swift trust in AI for the technology adaptation. An example items is "I trust that AI technologies can offer information and service that's best of my interest".

Procedure

For the validity and reliability, the measurement model has been established in which all items and related latent factors are included and linked to each other. The average variance extracted (AVE), composite reliability (CR), and Cronbach alpha coefficients (α) have been calculated. Then, structural equation model (SEM) was used for testing hypotheses and answering the research question (the fit indices used are $X^2/df < 3$, CFI and TLI > 0.90, and RMSEA < 0.08). The significance difference in countries has been calculated with *t*-test comparison. For all instruments, the five-point response category has been used ranging from "strongly disagree" (1) to "strongly agree" (5).

Results

Validity and reliability

The measurement model of the variables has been constructed by linking all research variables to each other. The measurement model has indicated the confirmation of the fit indices ($X^2/df = 2.46$, CFI = 0.96, TLI = 0.95, RMSEA = 0.047) after excluding an item from the swift trust scale that creates an inconsistency. Factor loadings of the variables have ranged from 0.55 to 0.88. The AVE and CR values for propensity trust are 0.54 and 0.82; human-like AI is 0.46 and 0.63; swift trust is 0.67 and 0.75; social influence is 0.64 and 0.73; functionality is 0.48 and 0.84, referring the construct and divergent validity. Cronbach alpha coefficients range from 0.72 to 0.86, indicating the internal consistencies.

Hypotheses testing

For testing the research hypotheses, we have constructed a SEM using the maximum likelihood estimation. In the SEM, the correlated variables of propensity trust, human-like AI, social influence, and functionality have been entered as exogenous and swift trust as endogenous variables. The structural model results have confirmed the fit indices (X^2/df = 3.02, CFI = 0.99, TLI = 0.99, RMSEA = 0.057). The standardized coefficients (beta) indicate that the influence of trust propensity on swift trust is 0.09 ($p < 0.05$, supporting *H1*), the influence of functionality on swift trust is −0.05 ($p > 0.05$, not supporting *H2*), the influence of social influence on swift trust is 0.25 ($p < 0.01$, supporting *H3*), and the influence of human-like AI on swift trust is 0.45 ($p < 0.01$, supporting *H4*). These results have indicated that the propensity of trust, social influence, and human-like AI have a positive relationship with swift trust, along with the effects of the functionality of AI.

Answering research question

To examine the research question, we divided the sample into two different samples of Polish and Turkish groups and calculated the multigroup analysis on the same SEM. The multigroup SEM analysis results have confirmed the fit indices (X^2/df = 2.19, CFI = 0.99, TLI = 0.99, RMSEA = 0.061). Based on the analysis results, there is a significant country-related difference in the effect of social influence on swift trust (Polish sample beta is 0.42, Turkish sample beta is 0.19, $t = 2.80$, $p < 0.01$), but not for the human-like AI on swift trust (Polish sample beta is 0.41, Turkish sample beta is 0.27, $t = 1.70$, $p > 0.05$). Findings have shown that Polish group has significantly higher levels of social influence on building swift trust in AI than Turkish group (Figure 8.1).

Conclusions

The purpose of the study was to explore the possible factors affecting trust in AI. The results present that propensity to trust significantly positively affects the development of swift trust in AI across the entire sample. This finding is consistent with the previously mentioned studies of Scholz et al. (2024).

The findings indicate that social norms also have a notable positive impact on the development of swift trust in AI throughout the entire sample. As previously mentioned, researchers hold differing viewpoints on the relationship between social influence and trust in AI. Our results are in some sense in line with the research results of Inaba and Togawa (2021). However, the positive impact of social norms is notably higher in the Polish group compared to the Turkish group.

Furthermore, the results indicate that the perception of human-like AI has a notable positive impact on the formation of swift trust in AI throughout the

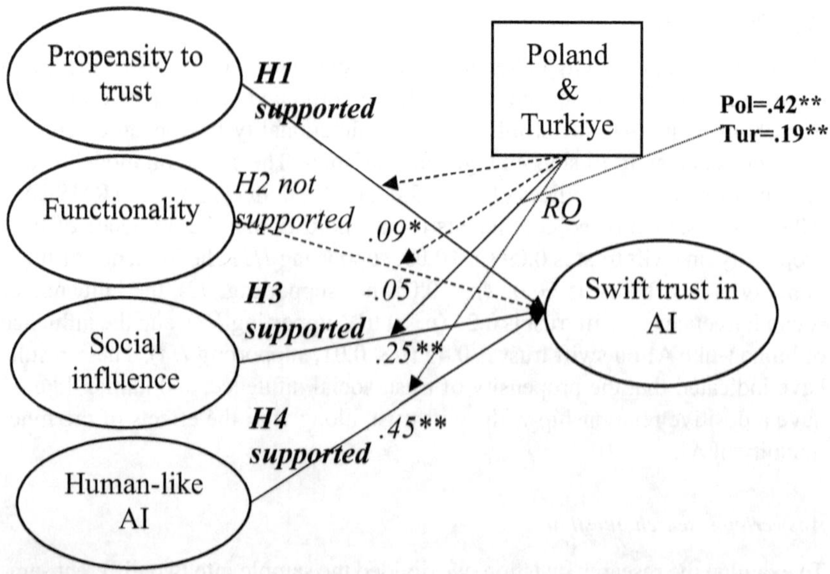

Figure 8.1 Research model and results
Source: Author's research.

entire sample. The results confirm earlier studies conducted by, among others, Roy and Naidoo (2021), Pelau et al. (2021), and Liu and Tao (2022), but on the other hand, they are contradictory to the study results of van Doorn Kim and McGill (2018) and Zang et al. (2021).

The research outcomes should be interpreted cautiously with the limitations of the cross-sectional research design, the survey methodology, the single-source data collecting, the generalizability derived from the student samples, and some of the statistics calculated just below the acceptable values.

In addition to enriching the existing literature, the research findings may also prove beneficial for AI developers and designers, organizations and businesses, and educators and trainers. Future research directions could help deepen the understanding of the complex dynamics between human-like AI, trust, and cultural or contextual factors. Further studies could examine how cultural contexts shape social norms and their effects on trust in AI across diverse regions or populations. Given the positive impact of human-like AI perception on swift trust, further research could investigate how specific anthropomorphic features (e.g., voice, appearance, emotional responsiveness) influence trust across different AI applications. While this study highlights the role of propensity to trust, subsequent research could also investigate individual differences in trust propensity across various AI contexts (e.g., health, education, public sector).

Acknowledgment

This research is partly funded by University of Economics Ho Chi Minh City, Vietnam.

References

Alarcon, G. M., Lyons, J. B., Christensen, J. C. et al. (2018). The effect of propensity to trust and perceptions of trustworthiness on trust behaviors in dyads. *Behavioral Research,* 50, 1906–1920.

Althuizen, N. (2018). Using structural technology acceptance models to segment intended users of a new technology: Propositions and an empirical illustration. *Information Systems Journal*, 28(5), 879–904.

Bach, T. A., Khan, A., Hallock, H., Beltrão, G., & Sousa, S. (2022). A systematic literature review of user trust in AI-enabled systems: An HCI perspective. *International Journal of Human-Computer Interaction*, 40(5), 1–16.

Barbul, M., & Bojescu, I. (2023). Generations' perception towards the interaction with AI. In *Conference: 9th BASIQ International Conference on New Trends in Sustainable Business and Consumption.* Constanța, Romania. https://doi.org/10.24818/BASIQ/2023/09/041

Choung, H., David, P., & Ross, A. (2022). Trust in AI and its role in the acceptance of AI technologies. *International Journal of Human-Computer Interaction*, 39(9), 1727–1739.

Deutsch, M., & Gerard, H. B. (1955). A study of normative and informational social influences upon individual judgment. *The Journal of Abnormal and Social Psychology*, 51(3), 629–636.

Epley, N., Waytz, A., & Cacioppo, J. T. (2007). On seeing human: A three-factor theory of anthropomorphism. *Psychology Review*, 114(4), 864–886.

Faulhaber, A. K., Ni, I., & Schmidt, L. (2021). The effect of explanations on trust in an assistance system for public transport users and the role of the propensity to trust. *Mensch Und Computer*, 303–310, https://doi.org/10.1145/3473856.3473886

Fishbein, M., & Ajzen, I. (1975). *Belief, Attitude, Intention, and Behavior: An Introduction to Theory and Research.*Reading, MA: Addison-Wesley.

Frazier, M. L., Johnson, P. D., & Fainshmidt, S. (2013). Development and validation of a propensity to trust scale. *Journal of Trust Research*, 3(2), 76–97.

Ganapati, S., & Desouza, K. (2025). Public value principles for secure and trusted AI. In *Proceedings of the 25th Annual International Conference on Digital Government Research* (pp. 251–257). Taipei, Taiwan. https://doi.org/10.1145/3657054.3657086

Inaba, Y., & Togawa, K. (2021) Social capital in the creation of AI perception. *Behaviormetrika*, 48, 79–102. https://doi.org/10.1007/s41237-020-00107-7

Jacquet, B., Baratgin, J., & Jamet, F. (2019). Cooperation in online conversations: The response times as a window into the cognition of language processing. *Frontiers in Psychology*, 10, 727.

Kim, H. Y., & McGill, A. L. (2018). Minions for the rich? Financial status changes how consumers see products with anthropomorphic features. *Journal of Consumer Research*, 45, 429–450. https://doi.org/10.1093/jcr/ucy006

Krafft, J., Twohig, M. P., & Levin, M. E. (2020). A randomized trial of acceptance and commitment therapy and traditional cognitive-behavioral therapy self-help books for social anxiety. *Cognitive Therapy and Research*, 44, 954–996.

Kramer, R. M. (1999). Trust and distrust in organizations: Emerging perspectives, enduring questions. *Annual Review Psychology*, 50(1), 569–598.

Küper, A., & Krämer, N. (2024). Psychological traits and appropriate reliance: Factors shaping trust in AI. *International Journal of Human-Computer Interaction*, 1–17. https://doi.org/10.1080/10447318.2024.2348216

Lee, J. D., & See, K. A. (2004). Trust in automation: Designing for appropriate reliance. *Human Factors*, 46(1), 50–80. https://doi.org/10.1518/hfes.46.1.50_30392.

Liu, K., & Tao, D. (2022). The roles of trust, personalization, loss of privacy, and anthropomorphism in public acceptance of smart healthcare services. *Computers in Human Behavior*, 127(5), 107026.

MacInnis, D. J., & Folkes, V. S. (2017). Humanizing brands: When brands seem to be like me, part of me, and in a relationship with me. *Journal of Consumer Psychology*, 27(3), 355–374.

Makiridakis, S. (2017). The forthcoming artificial intelligence (AI) revolution: Its impact on society and forms. *Futures*, 90, 46–60.

Maruping, L. M., Bala, H., Venkatesh, V., & Brown, S. A. (2017). Going beyond intention: Integrating behavioral expectation into the unified theory of acceptance and use of technology. *Journal of the Association for Information Science and Technology*, 68(3), 623–637. https://doi.org/10.1002/asi.23699

Mayer, R. C., Davis, J. H., & Schoorman, F. D. (1995). An integrative model of organizational trust. *The Academy of Management Review*, 20(3), 709–734. https://doi.org/10.2307/258792

Mayrashi, K. (2024). Determinants of public sector managers' intentions to adopt AI in the workplace. *International Journal of Public Administration in the Digital Age*, 11(1), 1–26.

McAllister, D. J. (1995). Affect- and cognition-based trust as foundations for interpersonal cooperation in organizations. *Academy of Management Journal*, 38(1), 24–59.

McKnight, D. H. (2005). Trust in information technology. In G. B. Davis (Ed.), *The Blackwell Encyclopedia of Management, Management Information Systems* (Vol. 7, pp. 329–331). Malden, MA: Blackwell.

McKnight, D. H., Carter, M., Thatcher, J. B., & Clay, P. F. (2011). Trust in a specific technology: An investigation of its components and measures. *ACM Transactions on Management Information Systems*, 2(12), 1–25.

McLean, G., & Osei-Frimpong, K. (2019). Hey Alexa… examine the variables influencing the use of artificial intelligent in-home voice assistants. *Computers in Human Behavior*, 99, 28–37. https://doi.org/10.1016/j.chb.2019.05.009

Moritz, J. (2017). Augmented humanity. *Technoetic Arts*, 15(3), 341–352.

Paliszkiewicz, J., & Gołuchowski, J. W. (2025). *Trust and Artificial Intelligence: Development and Application of AI Technology/Paliszkiewicz Joanna*, Gołuchowski Jerzy (*red.*), New York, Routledge, 3–9.

Pelau, C., Dabija, D. C., & Ene, I. (2021). What makes an AI device human-like? The role of interaction quality, empathy, and perceived psychological anthropomorphic characteristics on the acceptance of artificial intelligence in the service industry. *Computers in Human Behavior*, 106855. https://doi.org/10.1016/j.chb.2021.106855

Qin, F., Li, K., & Yan, J. (2020). Understanding user trust in artificial intelligence-based educational systems: Evidence from China. *British Journal of Educational Technology*, 51, 1693–1710.

Robert, L. P., Denis, A. R., & Hung, Y. C. (2009). Individual swift trust and knowledge-based trust in face-to-face and virtual team members. *Journal of Management Information Systems*, 26(2), 241–279.

Roy, R., & Naidoo, V. (2021). Enhancing chatbot effectiveness: The role of anthropomorphic conversational styles and time orientation. *Journal of Business Research*, 126(2021), 23–34.

Scholz, D. D., Kraus, J., & Miller, L. (2024). Measuring the propensity to trust in automated technology: Examining similarities to dispositional trust in other humans and validation of the PTT-A scale. *International Journal of Human-Computer Interaction*, 1–24. https://doi.org/10.1080/10447318.2024.2307691

Siau, K. L., & Wang, W. (2018). Building trust in artificial intelligence, machine learning, and robotics. *Cuttler Business Technology Journal*, 31(2), 48–53.

Tajfel, H. (Ed.). (1978). *Differentiation Between Social Groups: Studies in the Social Psychology of Intergroup Relations*. London: Academic Press Publishing.

Tinwell, A., Grimshaw, M., Nabi, D., & Williams, A. (2011). Facial expression of emotion and perception of the Uncanny Valley in virtual characters. *Computers in Human Behavior*, 27, 741–749. https://doi.org/10.1016/j.chb.2010.10.018

Todorov, A., Funk, F., & Olivola, C. Y. (2015). Response to Bonnefon et al.: Limited 'kernels of truth' in facial inferences. *Trends in Cognitive Sciences*, 19(8), 422–423.

van Doorn, J., Mende, M., Noble, S., Hulland, J., Ostrom, A., & Grewal, D. (2017). Domo Arigato Mr. Roboto: Emergence of automated social presence in organizational frontlines and customers' service experiences. *Journal of Service Research*, 20, 43–58. https://doi.org/10.1177/1094670516679272

Venkatesh, V., Morris, M. G., Davis, G. B., & Davis, F. D. (2003). User acceptance of information technology: Toward a unified view. *MIS Quarterly*, 27(3), 425–478.

Zang, S. Y., Meng, Z. X., Chen, B. B., Ywang, X., & Zhao, X. R. (2021). Motivation, social emotion, and the acceptance of artificial intelligence virtual assistants—trust-based mediating effects. *Frontiers in Psychology*, 12, https://doi.org/10.3389/fpsyg.2021.728495

Part III

Trust in specific contexts and applications

9 Trust in generative AI as seen by Gen Z in Albania, Poland, and Serbia

Iwona Pomianek, Etleva Muça, and Vesna Paraušić

Introduction

The evolution of Generative Artificial Intelligence (GenAI) has surprised the world and sparked both curiosity and delight in countless possibilities that until now have been treated by many as science fiction. GenAI is more than just a chatbot used for customer service and sending automated messages. It can generate innovative content, thereby creating innovations and improving the operation of systems in almost every industry. There are great expectations for GenAI – in the fields of health care (Zhang and Kamel Boulos 2023), education (Michel-Villarreal et al. 2023), art and creativity (Messer 2024), and development and innovations (Sedkaoui and Benaichouba 2024). However, entrepreneurs do not show an exceptionally high demand for AI tools. They are both curious and cautious, trying rather to determine the possible applications of GenAI in their businesses (How businesses... 2024). It raises the question of whether GenAI is capable of causing profound changes. History shows that technological breakthroughs, economy wide, take ages to pay off while the average worker needs time to get used to new ways of working. GenAI has enormous positive potential, and there is no doubt that such technologies can improve the lives of some and become a source of wealth for others. The technology is valuable and should probably not be stopped altogether. However, there must be appropriate regulations to ensure that both the development and use of GenAI do not lead to adverse harm (Sætra 2023).

Literature review

According to Polish sociologist Sztompka (2007), trust is a bet made about the uncertain future actions of other people and the confidence is the result of a certain calculation that it is worth taking the risk. According to Luhmann (2017), trust goes beyond information from the past and involves the risk of defining the future, thereby reducing the uncertainty of future events. The transfer of risk and trust attitudes from parents to offspring influences many aspects of children's

DOI: 10.4324/9781003586937-12

lives, potentially impacting their economic situation (Dohmen et al. 2012). People either have limited trust in innovative solutions, or their curiosity and willingness to test new solutions makes them take risks. Such modern solutions include AI and GenAI. Artificial Intelligence (AI) refers to the ability of computer systems to perform tasks that usually require human intelligence. This includes abilities such as learning, problem-solving, understanding natural language, and making decisions. AI relies on algorithms that analyze data, learn from it, and take action on it. AI systems are often used in process automation, image recognition, speech processing, and recommendation systems (Russel and Norvig 2009). Over the past ten years, the application of AI has gained a lot of interest in both academic and industry literature. AI is a young and developing topic. Numerous studies have examined how AI might enhance customer experience customization, optimize marketing strategies, and analyze vast volumes of data to derive insightful information (Toma and Hudea 2024; Utzi et al. 2025). GenAI is a field of AI that aims to generate new data based on existing data. These algorithms can create new content, such as text, images, sounds, or videos, based on patterns in training data. Examples include language models such as GPT and generative networks such as GAN. GenAI has applications in various fields, from artistic creation to 3D modeling to generating marketing content (Goodfellow et al. 2014; Radford et al. 2019; Chang and Kidman 2023). GenAI initially sparked resistance from the scientific community and debate over the ethics of using its tools by both scientists and students (Omrani et al. 2022; Amoozadeh et al. 2024). As recent research show (Gillespie et al. 2023), more than 60% people wary about trusting AI: it is the most trusted in health care, and the least trusted in human resources.

Generation Z (Gen Z) grew up in a world where digital technologies are integral to everyday life. As a result, they are more keen on adapting to new solutions, such as GenAI, and are more aware of their potential in both everyday life and professional work (Francis and Hoefe 2018; Puiu et al. 2022). That is why students will be more efficient and willing to use GenAI tools than lecturers, most often not worrying about ethical considerations more than necessary (for example, from the point of view of verifying semester papers or diploma theses in the anti-plagiarism system with the option of tracking GenAI participation) (Chan and Lee 2023; Ivanov et al. 2024).

Aim and methods

The authors set three research tasks. The first task was to present the views of Gen Z representatives (born between 1995 and 2012) in Albania, Poland, and Serbia on various issues related to GenAI tools and applications (apps). The views were examined regarding the following issues: knowledge of GenAI; types of GenAI apps used; trust in AI tools and frequency of using them; benefits of using GenAI and areas of greatest opportunity for using GenAI; barriers to using GenAI; the benefit to threat ratio of using GenAI; and evaluation of the level of user trust in GenAI in the coming years. The second research task was directed at examining the differences between respondents' attitudes according to the independent variables: country of origin (Albania, Poland, Serbia), gender (woman; man), professional status (employed; unemployed), and place of birth (village; small town; large town). The third research task included presenting the state of the analyzed field in all three countries involved in the study.

In order to address the tasks, the authors used desk research and survey research using a semi-structured online questionnaire. It contained eight questions about the studied area (with one or more answers offered) and questions about the socio-economic characteristics of the sample. Sampling was conducted in a purposeful-random way among representatives of Gen Z during 2024. The survey was conducted using Google Forms, which guaranteed the anonymity of the respondents. A total of 326 respondents participated in the study: 95 from Albania, 142 from Poland, and 89 from Serbia.

Descriptive statistics and $\chi 2$ test of independence were used to analyze and present the results. Since all tables were larger than 2×2, the assumption that at least 80% of cells have expected frequencies of five or more was respected in applying the $\chi 2$ test. When statistical significance of the association between the variables was found ($p \leq 0.05$), we used the Crosstabs procedure and the conclusion was made based on adjusted standardized residuals (Agresti 2002). The value of Cramer's V was used to estimate the effect size or strength of the association between the variables (Gravetter and Wallnau 2017).

Results and discussion

The survey research results show that many Gen Z representatives were familiar with AI. Namely, 83.7% of respondents confirmed whether they had heard of GenAI, while only 16.3% responded negatively. The $\chi 2$ independence test indicates no statistically significant differences in this response for any independent variable. As many as 284 respondents (87.1%) used only one AI app – ChatGPT. As for the frequency of using AI tools, the distribution of responses of the respondents who trusted AI tools was as follows: the most significant number of the respondents used AI tools rarely (63.4%), 26.9% of them used them frequently. In comparison, 9.7% of the respondents did not use them.

The responses mentioned above lead to the conclusion that Gen Z representatives in all three countries were well acquainted with GenAI and use AI apps and tools to a significant degree. Bearing in mind that 74% of the respondents were still studying or attending secondary education, we should assume that the respondents had a certain degree of caution when answering the question regarding the frequency of using AI tools (even though the survey was anonymous). It can be explained by the assumption that the respondents might be aware of a generally low level of approval of using ChatGPT in education from the teaching staff.

Gen Z puts excellent trust in AI tools. As many as 82.2% of the respondents (268) trusted these tools. At the same time, the respondents' views on this issue differ per country. Thus, Serbia is most trusted in AI tools (91.0%), while Poland is the least trusted (73.9%). Similarly, the data obtained by the χ^2 test of independence (Table 9.1) show that there are statistically significant differences between the respondents' views on this issue per country of origin ($\chi2$ (2, n = 326) = 12.443, p = 0.002). When evaluating the effect size based on the value of Cramer's V (0.2), the differences have small-to-medium effect. Using the Crosstabs procedure and interpreting the adjusted residuals, we can conclude that the percentage of respondents who expressed trust in AI was lower than expected for Poland and higher than expected for Serbia (where the expected values were those that would be achieved if the categorical variables were independent).

In addition to trust, significant AI issues are its limitations, risks, and threats on one hand and benefits and opportunities on the other. Examination of these issues shows the following:

- Regarding the most important benefits of using GenAI, the respondents were expected to select one or more of the following offered answers: creativity and innovation; personalization; efficiency and automation; decision support; cost reduction; improving user experience; predictive capabilities; possibilities of scaling results; and competitive advantage. For the most significant number of the respondents (46.3%), the most crucial benefit of GenAI was creativity and innovation (either as the only benefit or combined with other offered benefits). These are followed by 31.3% of the respondents who stated that the most important benefit of AI is efficiency and automation (either as the only benefit or combined with other offered benefits).
- More than half of the respondents, 57.4% (a total of 187), believed that the field with the most significant opportunities for using GenAI is data analysis, particularly when combined with another application field provided, such as company promotion; photo editing; automation of processes in the company; risk management in the company; creative design, e.g. logos and graphics; video generation; creating the content of letters in communication with the client; developing content for publication on the website; preparation of final papers at school/university; preparation of a diploma thesis; learning foreign languages; gaming; medical services; intelligent operation of electronic devices; or intelligent transport solutions.
- When it comes to the most important barriers to using AI tools, the largest percentage of the respondents (45.4%) stated that the most significant barrier was the quantity and quality of data available to GenAI, either as an independent barrier or combined with another barrier or multiple other barriers, such as possible bias (lack of objectivity) in the results generated by AI; high costs of data acquisition; interpretation difficulties; ethical issues; legal issues; quality control and verification; and security issues and lack of trust and reluctance of users.

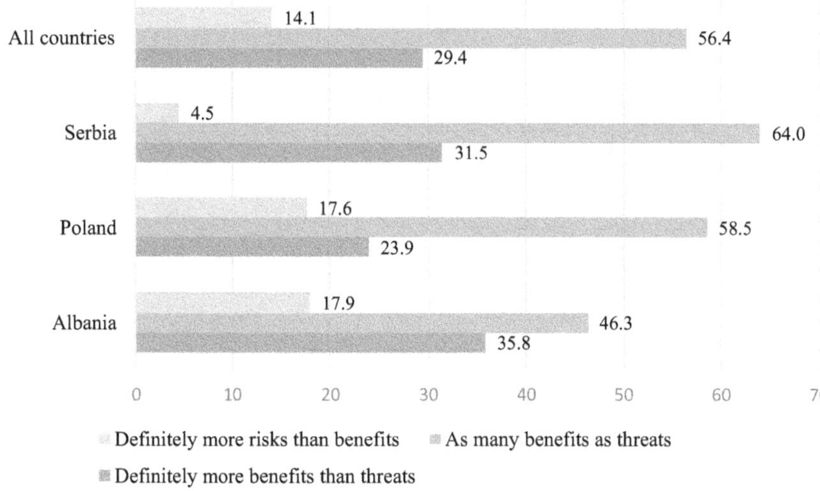

Figure 9.1 Benefit to threat ratio of using GenAI [% of respondents' answers]

Source: Authors' calculation

An interesting question related to assessing the benefit-to-threat ratio using GenAI was raised. The respondents were given the following answers: (1) more benefits than threats, (2) as many benefits as threats, and (3) more risks than benefits. The descriptive analysis shows that more than half of the respondents (56.4%) selected the answer "as many benefits as threats", which is a somewhat realistic and cautious standpoint (Figure 9.1). At the same time, a more significant percentage of the respondents believed that benefits outweigh threats (29.4%), compared to the group of respondents who believed that threats are more significant than benefits (14.1%) (Figure 9.1).

The data (Table 9.1) show that there are statistically significant differences between the respondents' views on this issue per country of origin ($\chi 2\ (4, n=326) = 13.646$, $p = 0.009$). When assessing the effect size of a country (as an independent variable) on the respondents' views on this issue, the value of Cramer's V (0.1) shows that these differences are of small effect. Interestingly, the application of the Crosstabs procedure and interpretation of the adjusted residuals show that the percentage of respondents who were skeptical about the benefits of AI is lower than expected in Serbia, compared to the other two countries. It does not mean that the respondents in Serbia evaluated AI tools more positively than those in Poland and Albania. It is more likely that the results indicate that a more significant percentage of the respondents in Serbia perceived AI as a tool with benefits and threats.

Finally, more than half of the respondents (174 respondents or 53.4%) believed that the level of user trust in GenAI would increase in the next five

Table 9.1 $\chi 2$ test of independence for examining the association among respondents' views on issues of AI and country of origin

Categorical variables	*Pearson chi-square, value**	*Df*	p
Do you trust AI tools? Offered answers: (1) yes; (2) no	12.443	2	0.002**
What is the benefit-to-threat ratio of using GenAI? Offered answers: (1) definitely more benefits than threats; (2) as many benefits as threats; (3) definitely more risks than benefits	13.646	4	0.009**
Do you think the level of user trust in GenAI will increase in the coming years? Offered answers: (1) yes, in the next five years; (2) yes, in the next ten years; (3) no, we still don't know GenAI's capabilities and the risk is high; (4) I have no opinion	16.761	6	0.010**

Source: Authors' calculation. The data were processed using the statistical software SPSS 25

*0 cells (0.0%) have expected count less than five

**Statistical significance at the alpha level of ≤0.05

years, while 73 respondents (or 22.4%) thought that this would happen in the following tenyears.

The $\chi 2$ test of independence (Table 9.1) indicates that there are statistically significant differences between the respondents' views on this issue per country of origin ($\chi 2$ (6, $n = 326$) = 16.761, $p = 0.01$). Based on the value of Cramer's V (0.2), these are the differences in medium effect size. Using the Crosstabs procedure and interpreting the adjusted residuals, we can conclude that in Serbia, compared to the other two countries, the percentage of the respondents who believed that the level of user trust in GenAI would increase in the next five years is higher than expected, while the percentage of those who were skeptical about AI is lower than expected. It indicates that the respondents' views were generally more positive in Serbia than in the other two countries. At the same time, when it comes to the respondents who still had no opinions about this issue, the adjusted residuals show that the percentage of these respondents was higher than expected for Albania and lower than expected for Poland. It can be suggested that Gen Z representatives in Albania were still quite uncertain about how user trust in GenAI would develop in the coming period.

AI is a new concept in Albania. Many young people see AI as the savior of their time, for their daily tasks. Digital transformation in Albania was more developed due to the pandemic that changed the way of thinking globally (Boce and Myrto 2022). According to Karafili et al. (2022), digital skills and AI are necessary for all graduates after high school, to increase their professional competencies. This process requires the development of curricula for students. The cornerstone for the growth of intelligent education in Albania will be laid by the total execution of the development policies, the National Strategies, and the notable augmentation of the government's investments in the education sector and the sectors that bolster it (Kraja 2023).

By the end of 2024, the National Information Society Agency should finalize the National Strategy for Artificial Intelligence for the first time. This strategy would focus on four main issues that will be considered (Government of Albania Policy Document 2024):

- increasing access to digital technologies, both in terms of geographical penetration and financial affordability;
- digital literacy, not only for students, young people and professionals in the sector, but also for the adult population;
- protection of personal data by European Union standards;
- cybersecurity in coordination with our strategic Euro-Atlantic partners.

On the other hand, Albanian companies are still waiting to adopt AI in their daily processes, as they believe it will reduce human capital and increase their costs (Kaso and Xhindi 2023). Among young students, AI focuses more on social media, such as Facebook, Instagram, and more (Pomianek and Muça 2024).

The document that sets out the goals and actions in Poland in the field of AI is the "Policy for the development of artificial intelligence in Poland from 2020" (Policy… 2020). The document covers six areas: society, innovative companies, science, education, international cooperation, and the public sector. In Poland, AI is still not very popular. As studies show (Brylska et al. 2024; Nowina 2023), Poles are beginning to understand the role of AI, but they still have many questions and concerns about it. With the further dissemination of AI in everyday life, changes in how it is perceived can be expected.

Serbia was the first country in Southeast Europe to strategically regulate AI in 2019 to enhance its safe use by international ethical regulations (Strategy Serbia 2019). The most significant achievements in AI development in Serbia in the previous period have been (1) the establishment of the National Institute for AI development ("Research and Development Institute for AI of Serbia"); (2) introduction of educational Master's programs for AI at three universities; and (3) establishment of the National AI Platform (supercomputer), which is located in the State Data Centre in Kragujevac (Action Plan 2020). Access to the National AI Platform and use of the supercomputer is free for universities, scientific institutes, faculties, and domestic start-up companies in science and technology parks.

In education, the mandatory curriculum in primary and secondary schools in Serbia involves the field of AI. In addition, there is an increasing number of

Master's programs related to AI at universities (National AI Platform – Serbia). The literature review shows that the application of AI has a positive impact on the efficiency of the teaching process and the overall educational system in Serbia, and that students have a positive attitude toward the application of AI (Ilić et al. 2021; Kuleto et al. 2022; Tomić and Radovanović 2024). Students believe that new technologies (AI, machine learning, extended reality) help develop their skills, creating an encouraging research environment and space for collaborative learning, and that these technologies increase their motivation for learning and working together with other students (Ilić et al. 2021). Similar results have been reached by the group of authors (Kuleto et al. 2022), who analyzed the opinions of students in Serbia, Romania, and Portugal about the implementation of blockchain technology. Their findings indicate that students think this technology positively impacts learning performance and can improve trust and transparency in higher education (Kuleto et al. 2022).

Serbia is accelerating the progress in the field of AI development. In 2022, the country joined the Global Partnership on AI, while in 2023, the Serbian government adopted ethics guidelines for the development of AI. In 2024, the Council for AI was established. A new strategic document on AI development for 2024–2030 is being adopted, while laws and bylaws in this field are being prepared.

Conclusions

Having grown up in the digital era, Gen Z in Albania, Poland, and Serbia is typically more used to technology, especially AI. Because GenAI technologies are widely regarded as innovative and helpful for various applications, including content production and problem-solving, this familiarity may increase people's trust. Gen Z may become aware of the potential benefits of GenAI as it permeates everyday life through social media, entertainment, and educational materials. The results of this chapter, conducted in three countries, indicate that policymakers and educational organizations must enhance their curricula to emphasize the significance of emerging technologies. This conclusion is evident from other authors studying the future necessities of Gen Z (Vitezić and Perić 2021). AI may have unfavorable effects on educational standards and the standard of teaching. Following ChatGPT's introduction, educational institutions all around the world came up with solutions to address these issues.

Gen Z may acknowledge its potential advantages as GenAI becomes increasingly embedded in daily life – via social media, entertainment, and educational resources. Gen Z mainly obtains information and influence via social media channels. Conversations, patterns, and experiences within their networks can significantly impact how they see GenAI. Favorable recommendations or unfavorable encounters influence their trust levels.

References

Action Plan (2020). Action Plan for 2020–2022 for the Implementation of the Strategy for the Development of AI in the Republic of Serbia for the Period 2020–2025. Official Gazette of the RS, No. 81/20. Available at: https://nitra.gov.rs/images/ministarstvo/dokumenta/akcioni-plan-strategija-vestacke-inteligencije.pdf

Agresti, A. (2002). *Categorical Data Analysis*. New York: Wiley. Available at: https://lin-yu.me/books/AGRESTI_categorical_data_analysis_2nd.pdf

Amoozadeh, M., Daniels, D., Nam, D., Kumar, A., Chen, S., Hilton, M., Ragavan, S.S., Alipour, M.A. (2024). Trust in generative AI among students: An exploratory study. In *SIGCSE 2024: Proceedings of the 55th ACM Technical Symposium on Computer Science Education*, Vol. 1, pp. 67–73. https://doi.org/10.1145/3626252.3630842

Boce, G., Myrto, E. (2022). Transformimi Dixhital dhe Roli në Teknollogjinë e Informacionit. *Euro Mediterraean Journal, 17*, 88–97.

Brylska, K., Balcerzak, A., Gackowski, T., Kowalczyk, G., Łączyński, M., Mierzecka, A., Sztyber, M., Blak, A., Hykawy, J., Szydelski, J., Sarna, M., Zakrzewska, J. (2024). *Poles on Artificial Intelligence: CATI Survey Report by LBM UW*. Laboratorium Badań Medioznawczych UW. Available at: https://www.lbm.uw.edu.pl/images/Raport_AI_ENG2.pdf

Chan, C.K.Y., Lee, K.K.W. (2023). The AI generation gap: Are Gen Z students more interested in adopting generative AI such as ChatGPT in teaching and learning than their Gen X and millennial generation teachers?. *Smart Learning Environments, 10*, 60. https://doi.org/10.1186/s40561-023-00269-3

Chang, C.H., Kidman, G. (2023). The rise of generative artificial intelligence (AI) language models - challenges and opportunities for geographical and environmental education. *International Research in Geographical and Environmental Education, 32*(2), 85–89. https://doi.org/10.1080/10382046.2023.2194036

Dohmen, T., Falk, A., Huffman, D., Sunde, U. (2012). The intergenerational transmission of risk and trust attitudes. *The Review of Economic Studies, 79*(2), 645–677. https://doi.org/10.1093/restud/rdr027

Francis, T., Hoefe, F. (2018). True Gen: Generation Z and Its Implications for Companies. *McKinsey & Company*. Available at: www.mckinsey.com/~/media/McKinsey/Industries/Consumer%20Packaged%20Goods/Our%20Insights/True%20Gen%20Generation%20Z%20and%20its%20implications%20for%20companies/Generation-Z-and-its-implication-for-companies.pdf

Gillespie, N., Lockey, S., Curtis, C., Pool, J., Akbari, A. (2023). Trust in Artificial Intelligence: A Global Study. *The University of Queensland and KPMG Australia*. https://doi.org/10.14264/00d3c94

Goodfellow, I.J., Pouget-Abadie, J., Mirza, M., Xu, B., Warde-Farley, D., Ozair, S., Courville, A., Bengio, Y. (2014). Generative Adversarial Networks. https://doi.org/10.48550/arXiv.1406.2661

Government of Albania (2024). Priorities Policy Document. Decision no. 533, dated 12.8.2024. Available at: https://faolex.fao.org/docs/pdf/alb229673.pdf

Gravetter, F., Wallnau, L. (2017). *Statistics for the Behavioral Sciences*, 10th edition. Boston: Cengage Learning.

How Businesses Are Actually Using Generative AI (2024). *The Economist*, February 29. Available at: https://www.economist.com/business/2024/02/29/how-businesses-are-actually-using-generative-ai

Ilić, M.P., Paun, D., Popovic Ševic, N., Hadžic, A., Jianu, A. (2021). Needs and performance analysis for changes in higher education and implementation of artificial intelligence, machine learning, and extended reality. *Education Sciences, 11,* 568. https://doi.org/10.3390/educsci11100568

Ivanov, S., Soliman, M., Tuomi, A., Alkathiri, N.A., Al-Alawi, A.N. (2024). Drivers of generative AI adoption in higher education through the lens of the theory of planned behaviour. *Technology in Society*, *77*, 102521. https://doi.org/10.1016/j.techsoc.2024.102521

Karafili, Z.E., Nientied, Z.P., Aliaj, Z.B. (2022). Evaluation of the Regulation Framework Cadre of the Professional and High Education in Albania. UNDP Report. Available at: https://www.undp.org/sites/g/files/zskgke326/files/2022-05/Review%20of%20the%20Regulatory%20Framework%20and%20Provision%20of%20Post-Secondary%20and%20Higher%20VET%20in%20Albania.pdf

Kaso, E., Xhindi, T. (2023). The use of artificial intelligence and machine learning technology by companies in Albania. In *Challenges of the Society Facing, Book of Proceedings*, pp. 203–218. Available at: https://www.umsh.edu.al/media/650d6a5695eea.pdf

Kraja, P. (2023). Report on smart education in Albania. In Zhuang, R., Liu, D., Sampson, D., Mandic, D., Zou, S., Huang, Y. (Eds.), *Ronghuai Huang Smart Education in China and Central & Eastern European Countries* (pp. 51–79). Singapore: Springer Nature.

Kuleto, V., Bucea-Manea-Ţoniş, R., Bucea-Manea-Ţoniş, R., Ilić, M.P., Martins, O.M., Ranković, M., & Coelho, A.S. (2022). The potential of blockchain technology in higher education as perceived by students in Serbia, Romania, and Portugal. *Sustainability, 14*(2), 749. https://doi.org/10.3390/su14020749

Luhmann, N. (2017). *Trust and Power*. Cambridge: Polity. Available at: https://www.wiley.com/en-us/Trust+and+Power-p-9781509519453

Messer, U. (2024). Co-creating art with generative artificial intelligence: Implications for artworks and artists. *Computers in Human Behavior: Artificial Humans*, *2*(1), 100056, https://doi.org/10.1016/j.chbah.2024.100056

Michel-Villarreal, R., Vilalta-Perdomo, E., Salinas-Navarro, D.E., Thierry-Aguilera, R., Gerardou, F.S. (2023). Challenges and opportunities of Generative AI for higher education as explained by ChatGPT. *Education Sciences, 13,* 856. https://doi.org/10.3390/educsci13090856

National AI Platform - Serbia. Available at: https://www.ai.gov.rs/tekst/en/95/national-ai-platform.php#

Nowina Konopka, M. (2023). Każdy zna się na AI: przegląd badań polskiej opinii publicznej na temat sztucznej inteligencji. *Zeszyty Prasoznawcze, 66*(4(256)), 127–142. https://doi.org/10.4467/22996362PZ.23.043.18677

Omrani, N., Rivieccio, G., Fiore, U., Schiavone, F., Garcia Agreda, S. (2022). To trust or not to trust? An assessment of trust in AI-based systems: Concerns, ethics and contexts. *Technological Forecasting and Social Change, 181,* 121763, https://doi.org/10.1016/j.techfore.2022.121763

Policy (2020). Policy for the Development of Artificial Intelligence in Poland from 2020. Available at: https://www.gov.pl/attachment/928200fa-b1a6-4c0c-b3a8-d1fbf1e1175a

Pomianek, I., Muça, E. (2024). New dimensions of the social capital affected by digital communication. In Paliszkiewicz, J., Chen, K., Mendel, M. (Eds.), *Trust in Social and Business Relations: Theory and Practice*, pp. 251–261. New York: Routledge.

Puiu, S., Demyen, S., Tănase, A.-C., Vărzaru, A.A., Bocean, C.G. (2022). Assessing the adoption of mobile technology for commerce by Generation Z. *Electronics, 11,* 866. https://doi.org/10.3390/electronics11060866

Radford, A., Wu, J., Child, R., Luan, D., Amodei, D., Sutskever, I. (2019). Language models are unsupervised multitask learners. *OpenAI blog*, *1*(8), 9. Available at: https://www.bibsonomy.org/bibtex/61ea7e007d6c95171a2ff3396b1af7d9

Russel, S.J., Norvig, P. (2009). *Artificial Intelligence: A Modern Approach.* Upper Saddle River: Pearson Education Inc. Available at: https://people.engr.tamu.edu/guni/csce421/files/AI_Russell_Norvig.pdf

Sætra, H.S. (2023). Generative AI: Here to stay, but for good? *Technology in Society, 75*, 102372, https://doi.org/10.1016/j.techsoc.2023.102372

Sedkaoui, S., Benaichouba, R. (2024). Generative AI as a transformative force for innovation: A review of opportunities, applications and challenges. *European Journal of Innovation Management*, ahead-of-print. https://doi.org/10.1108/EJIM-02-2024-0129

Strategy Serbia (2019). Strategy for the Development of AI in the Republic of Serbia for the period 2020-2025. Official Gazette of the RS, No. 96/19. Available at: https://www.media.srbija.gov.rs/medsrp/dokumenti/strategy_artificial_intelligence.pdf

Sztompka, P. (2007). *Zaufanie. Fundament społeczeństwa*. Kraków: Znak.

Toma, S.G., Hudea, O.S. (2024). Generation Z students 'perceptions on the abilities, skills and competencies required in the age of artificial intelligence systems. *Amfiteatru Economic, 26*(65), 162–180.

Tomić, B.M., Radovanović, N.D. (2024). The application of artificial intelligence in the context of the educational system in Serbia, with a special focus on religious education. *Sociološki pregled, 58*(2), 435–459. https://doi.org/10.5937/socpreg58-48911

Utzi, M., Kapaj Mane, A., Raczkowska, M. (2025). The role of communication in building trust in e-commerce from the perspective of consumers. In Paliszkiewicz, J., Kuanchin, C., Mendel, M. (Eds.), *Trust in Social and Business Relations: Theory and Practice*, pp. 204–212. New York: Routledge.

Vitezić, V., Perić, M. (2021). Artificial inteligence acceptance in services: Connecting with Generation Z. *The Service Industries Journal, 41*(13–14), 926–946. https://doi.org/10.1080/02642069.2021.1974406

Zhang, P., Kamel Boulos, M.N. (2023). Generative AI in Medicine and healthcare: Promises, opportunities and challenges. *Future Internet*, *15*, 286. https://doi.org/10.3390/fi15090286

10 Cultivating confidence

The intersection of artificial intelligence and trust in commercial aviation

Leila Halawi, Mark Miller, and Sam Holley

Introduction

Artificial intelligence (AI) is pervasive in scholarly publications, internet sites, and public discourse. AI is a term with broad scope that refers to machines that can learn and perform tasks that typically require human intelligence. The specter of AI intruding into many aspects of aviation has raised alarms, concerns, and prodigious misunderstanding of potential and contemplated applications in systems and processes. The EASA AI Roadmap (EASA, 2023), a linear projection with three levels extending into 2050, places the human-AI teaming period (through 2035) at Level 2. This suggests a ten-year span to develop the interactive issues to accomplish reliable and transparent generative AI in aviation flight operations. To aid in bifurcating some of the current angst regarding popular conceptions of how generative AI will be increasingly invasive in systems, a proposed twin trajectory approach has been suggested (Holley et al., 2024), which follows a continuum of automation evolution that interacts with a domain-specific approach for current and near-current applications and with a more complex approach that addresses human cognition and machine testing.

The authors envision trajectories showing a progression or roadmap of AI technologies across two main categories: (1) complex AI and (2) domain-specific AI. Each category has a series of concepts or technologies arranged linearly, suggesting an evolution or increasing complexity. The pathways start with more fundamental or specific technologies and progress toward more advanced, integrated, and generalized AI systems. For instance, the complex AI path moves from Augmented Cognition through CAPS/CHARM and Autonomous Operations to Generative AI Neurospace. The Automation path begins with basic Automation and progresses through stages like Cognitive Echelons and Machine Learning, ultimately leading to Adaptive Allocation. The paths converge, implying that these AI approaches may eventually integrate or lead to similar advanced outcomes in AI development. The domain-specific AI path starts with more specialized systems. It progresses toward more advanced and adaptable technologies, including Auto-pilot GPWS TCAS, GPWS (Ground

DOI: 10.4324/9781003586937-13

Proximity Warning System), and TCAS (Traffic Collision Avoidance System), Augmented Automation, suggesting enhancing automated systems with additional capabilities, possibly integrating more data or decision-making abilities; Adaptive Automation implying systems that can adjust their behavior based on changing conditions or requirements, Future Systems, a broader term suggesting more advanced AI systems that build upon the previous stages. Lastly, adaptive allocation, the final stage, shared with the automation path, likely refers to AI systems capable of dynamically allocating resources or making decisions across various domains. This progression shows a path from highly specific, rule-based systems to more flexible and broadly applicable AI technologies in specialized domains (Miller et al., 2023; Miller et al., 2024).

Critical issues of reliability, transparency, and trust are at the core of this evolution, particularly concerning the implementation of AI systems. As the commercial aviation industry cautiously explores AI applications, understanding public perception and acceptance becomes paramount. Of particular interest is how different generational cohorts with distinct technological experiences and attitudes may vary in their trust of AI in aviation contexts. This study investigates the intersection of AI trust and generational attitudes within the promising field of AI adoption in US commercial aviation. By examining the unique characteristics and technological inclinations of various age groups, we seek to uncover insights that could inform the development and implementation of AI systems in aviation, ultimately contributing to this critical industry's safe and accepted progression.

Literature review

Function allocation: human-robot/AI teaming

In the realm of advanced AI, machines capable of learning and performing tasks typically requiring human intelligence are redefining human-machine interactions in aviation. The current focus is Level 1 AI (EASA, 2023), introducing intelligent assistants for specific functions. Progressing to Level 2 involves knowledge capture from deep learning, demanding assurance of representative data encompassing situational variabilities. Like Daedalean's neural network vision-based traffic detection system, domain-specific functions build operator confidence as regular assistive functions (Warwick, 2024). Reliability and trustworthiness become more elusive for AI emulating human cognition, partly due to function allocation (Hancock, 2023). This involves determining how functions are distributed between humans and machines and who performs and authorizes the allocation. These considerations raise concerns about situational comprehension and judgment. Trust-building through experiential learning is crucial in human-machine teaming. Level 3 considerations involve cognitive-cyber symbiosis, allowing autonomous relationships to develop cautiously (Abbass, 2019).

Challenges in aviation AI include data management, scalability, safety, explainable AI, rigorous validation, avoiding unintended consequences, and addressing data distribution shifts (Brooker, 2024). Future systems may face obstacles when integrating AI, such as in digital flight rules (Wing et al., 2022). Interactions with air traffic control and operations centers further complicate these issues. The development of AI in aviation is branching into a complex track oriented toward cognitive interactions and potential teaming with machine intelligence. Trust remains a primary concern in human-AI teaming, emphasizing the necessity for human participation in the decision loop (Nørgaard & Linden-Vørnle, 2021).

Confidence & robotics

As human-robot collaboration (cobotics) evolves, confidence in each party's accuracy becomes crucial for safety and effectiveness, especially in complex systems. Determining acceptable risk thresholds raises questions about ethics, liability, and legal parameters. Confidence in AI decisions relates to boundary conditions for perceptions and fairness in decision-making. Araujo et al. (2020) studied "algorithmic appreciation," where AI recommendations are preferred over human ones, particularly when compared to expert decisions. However, this preference does not apply to non-experts. Risk perception associated with AI decisions influences human decision-making, as humans are more likely to consider these factors. Johnson and Vera (2019) emphasize the importance of reconciling these conditions to increase acceptable confidence thresholds in cobotic relations. Balancing AI capabilities with human expertise and risk awareness is essential for developing effective and trustworthy cobotic systems.

Present systems for structuring human-computer conversations rely on decision trees, which require developers to implement all possible communication paths. This approach leads to complicated, difficult-to-maintain node networks and often results in awkward interactions, unlike natural human conversations. Current systems also fail to utilize analytics to guide conversation flow or apply information from previous interactions (Woodard & Fairbrother, 2020). It is important to differentiate between neurotechnology, which deals with human cognitive skills and brain activity, and complex AI, which attempts to mimic human brain function using algorithms, sensors, and robotic systems to create neural networks. In AI, human neurons are replaced by network neurons performing functions based on mathematical models that collect and classify information according to specific architectures. Systems coding neural patterns of human operators involve an increasing understanding of neuroscience, advancements in neuroimaging, and cyber-microsystems (Schubert et al., 2023). As AI capabilities evolve beyond set functions and efforts to create human-AI hybrid systems progress, the unpredictable nature of these developments poses challenges for ensuring reliability. Integrating neurotechnology and AI in creating more natural and efficient human-computer interactions represents a frontier in technological development, with significant potential for improving user experiences while raising important questions about system reliability and the nature of human-machine collaboration.

AI in commercial aviation offers advantages like real-time data processing, cognitive load reduction, and increased autonomy. AI enhances commander decisions in military applications but raises accountability concerns in neurospace operations. The shift to distributed network models recognizes collaborative risk functions between humans and technology, fostering advanced human-machine symbiosis. Neurocognitive network systems drive rapid advances in AI coding, informed by human neural system studies. This enables the integration of human and AI cognitive strengths in hybrid systems. However, human involvement remains crucial for balanced judgment and sensitive protocols (Nørgaard & Linden-Vørnle, 2021). The evolution of these systems facilitates multi-domain operations, including collaboration between robotic and autonomous systems, while acknowledging the need for human oversight in decision-making processes.

Adaptive neurofeedback systems could enhance cross-domain synergy in time-sensitive environments. Closed-loop cognition may mediate machine-human perception, allowing adaptation through iterations. This allows the human operator to respond to unforeseen changes and exercise discretionary judgment. Challenges include trust, transparency, and preventing algorithmic errors. The responsibility gap in conjoined human-robot operations requires resolution for progress. A better understanding of closed-loop systems is needed to avoid cognitive overload in operators. This collaboration could enable human operators to respond to unforeseen changes and exercise discretionary judgment. However, issues like algorithm cheating and counterintuitive results must be addressed. Resolving these challenges is crucial for significant advancements in human-machine interaction.

Trust

McKnight et al. (2002) identified that the research trajectories concerning technology trust predominantly focused on the human dimensions and characteristics contributing to the perception of trustworthiness, thereby fostering a user's confidence and acceptance of technology. Thatcher et al. (2011) showed that perceived usefulness and ease of use significantly influenced trust in technology regarding the intention to use innovative knowledge management system applications. Their findings greatly impacted subsequent research frameworks and methodologies in the field (McKnight et al., 2011; Thielsch et al., 2018, and Söllner, 2020). For a deeper understanding of the concept of trust, we suggest reviewing the work by Halawi et al. (2024), which employs the technology acceptance model to investigate perceptions of AI in aviation. The study unveiled intricate interrelations between perceived utility, trust, and adoption intentions. These results accentuate the complex dynamics inherent in trust formation concerning AI integration in aviation and highlight its crucial significance for effective implementation.

Generational differences

The Baby Boomer generation, born between 1945 and 1964 (Harlianto & Rudi Cheulie, 2020), faces a complex decision-making landscape regarding retirement, influenced by increased longevity and diverse financial options (Altman et al., 2020). This cohort and Generation X exhibit a slower adoption rate for new technologies than their younger counterparts (Nakagawa & Yellowlees, 2020). Millennials, born between 1980 and 2000 (Yadav & Chaudari, 2024), demonstrate superior multitasking abilities and a greater reliance on technology, attributed to their immersion in a digitally saturated environment from an early age (Nakagawa & Yellowlees, 2020). This technological affinity may significantly influence their trust in AI systems within the aviation industry. Each generational cohort possesses unique attitudes, educational aspirations, and work performance outcomes, which may impact their perception and acceptance of AI in aviation contexts. Lazanyi (2019) studied young generations' attitudes toward AI and robots. Gen Z showed more willingness to engage with robots than Gen Y, driven by a desire for personal contact rather than loneliness. Younger generations see tech as enhancing, not replacing, personal connections. According to Borkovich (2014), digital natives and older groups trust online news similarly, challenging assumptions about younger generations. Personalization and echo chambers influence trust across ages. The varying levels of work engagement across generations could correlate with their willingness to embrace AI-driven systems in aircraft operations.

The evolution of AI: balancing trust across generations

Commercial aviation has evolved significantly over the past 50 years, transitioning from traditional safety approaches to a collaborative risk management mindset embodied in Safety Management Systems (SMS). These systems integrate human factors tools like Crew Resource Management (CRM) and Threat and Error Management (TEM) with advanced flight safety technologies (FAA, 2014). The industry now stands at a crossroads, with the rapid growth of AI promising a new era of proactive, data-driven SMS. Simple AI systems have already significantly impacted flight operations. GPWS and Traffic Alert Collision Avoidance Systems (TCAS) exemplify how AI has addressed critical safety hazards like Controlled Flight Into Terrain and mid-air collisions. These innovations demonstrate successful human-machine teaming, enhancing pilot decision-making and aviation safety over the last three decades.

Pilots have increased trust in AI decision-making tools through constant improvement and standardized training. However, generational differences affect technology trust and situational awareness. Senior pilots might view EGPWS (Enhanced Ground Proximity Warning System) or Terrain Awareness Warning System (TAWS) as an extra safety tool for maintaining high terrain awareness. In contrast, younger Generation Z or A pilots, who grew up with immersive computer games, may over-rely on these systems, trusting them more than maintaining detailed situational awareness. This generational mindset difference is crucial as AI becomes more involved in various aviation operations, including air traffic control, maintenance, ramp, and dispatch.

Trust in AI must be paramount in commercial aviation, with reliable systems and proper human-machine teaming. Understanding generational differences in AI interaction could significantly impact training design for improved decision-making. Kolmos (2017) found that pilots over 40 have more difficulty transitioning to computerized flight decks and are more prone to accidents than their younger counterparts. As aviation technology advances toward automation and AI, addressing these generational gaps becomes crucial for maintaining safety standards.

As aviation progresses, AI will enhance decision-making capabilities by collecting vast amounts of data, especially human factors-related information, for proactive data-driven SMS. Managers must develop appropriate trust in AI for effective teaming decisions, similar to how pilots learned to trust EGPWS/TAWS and TCAS.AI teaming has significant potential in commercial aviation safety, enhancing SMS by merging multiple data streams to complement risk management and hazard identification with human factors root causes. This evolution parallels the development of CRM and TEM in flight operations. As safety data collection tools evolve with AI, managers must trust AI-assisted decision-making. The risk associated with management and flight operations decisions necessitates a perfect balance of trust, achieved through proper training for managers and supervisors. All generations (Boomers, X, Millennials, Z, and Alphas) will initially have different predisposed attitudes and trust factors toward AI teaming in aviation safety decisions. The key factor will be finding the proper balance of trust between humans and AI to optimize decision-making in aviation safety management.

Methodology

This research explored the relationship between age and trust in AI within the US commercial aviation industry. Following IRB approval, participants were recruited from the US commercial aviation sector, with the inclusion criterion that respondents must be US citizens engaged in commercial aviation. SurveyMonkey was used for questionnaire dissemination and data collection across diverse geographical locations. The survey included anonymous demographic and screening questions.

Trust in AI was assessed using a validated scale, demonstrating reliability in previous research (Halawi et al., 2024; McKnight et al., 2002). This scale was adapted for the aviation context, employing a 7-point Likert scale from 1 (strongly distrust) to 7 (firmly trust). Demographic data collected included age, gender, education, aviation experience, and industry roles. Age data were categorized into five groups: <18, 18–29, 30–44, 45–60, and >60 years old.

Over six months, 273 eligible responses were collected. Normality tests were conducted to ensure the appropriateness of parametric statistical methods (Halawi et al., 2024). Trust was operationalized using a nine-item scale based on McKnight et al. (2002), refined through factor analysis in SPSS, yielding an eight-item, single-factor solution with high construct validity (all loadings > 0.5) and excellent reliability (Cronbach's $\alpha > 0.934$) (Halawi et al., 2024). A one-way ANOVA was employed to examine differences in trust across age groups. The independent variable was the age group, while the mean trust score was the dependent variable. SPSS version 28 was used for analysis. Descriptive statistics were computed for each age group, including means, standard deviations, and 95% confidence intervals. The effect size was assessed using eta-squared to evaluate practical significance.

The study tested the following hypothesis:

Null hypothesis (H0): There is no significant difference in perceptions of AI use and trust across different generations within the US commercial aviation industry.

Alternative hypothesis (H1): A significant difference exists in perceptions of AI use and trust across different US commercial aviation industry generations.

Results and discussion

In our quest to understand how different generations perceive and trust AI in aviation, we conducted a comprehensive analysis that has yielded some intriguing insights. Our research, which encompassed a variety of age demographics, sought to ascertain whether chronological age significantly influences trust toward AI in the aviation industry. Our survey encompassed a heterogeneous cohort of 273 participants, ranging from young adults to older individuals. We solicited their evaluations of trust in AI within the aviation sector on a scale from 1 to 7, where elevated scores signified an augmented level of trust. Our investigation examined the variations in trust toward AI within aviation across distinct age demographics. The youngest demographic in our analysis, aged 18 to 29, demonstrated moderate trust, with a mean score of approximately 3.9 out of 7. This implies a prudent yet not wholly skeptical stance toward AI in aviation among young adults. As we progress to the next age bracket, a slight decline in trust is evident among participants aged 30 to 44, with their average trust score decreasing marginally to 3.8. This group, representing the most substantial segment of our respondents, exhibits the greatest caution regarding AI in aviation. Their mid-career status and potential frequency of air travel may render them more critical of emerging technologies in the aviation sector. The trend shifts positively when we examine the older age groups. Individuals aged 45 to 60 displayed heightened trust, with their average score ascending to approximately 4.1. However, the most astonishing revelation emerged from our oldest demographic, those exceeding 60 years of age. They manifested the highest trust in AI for aviation, achieving an average score nearing 4.5. This trend contradicts the stereotype that older populations are inherently more resistant to novel technologies. Within the context of aviation AI, our senior participants appear to exhibit the greatest openness and trust.

Our research challenges common assumptions about age and technology acceptance, revealing a more complex relationship. While teens often exhibit tech expertise from leisure activities (Johnson, 2009), trust in aviation AI surprisingly increases marginally with age. This finding suggests that domain expertise and experience may be more influential than generational factors in shaping attitudes toward AI in specialized fields. Statistical analysis supports these conclusions. With an F-value of 2.532 and 4 degrees of freedom between groups (five age cohorts), the Mean Square Between Groups (4.798) exceeds the Mean Square Within Groups (1.895). The p-value of 0.041 indicates statistically significant differences in AI trust among age groups. However, of the total variability (Sum of Squares Total: 526.977), only 19.192 is attributed to age group differences, while 507.785 is due to intra-group variability. These findings caution against broad generalizations about technology acceptance based solely on age, especially in specialized fields like aviation. Statistical analysis reveals significant variations in AI trust across age groups in aviation, though age's influence is minimal. Only 3.6% of total variability is attributed to age differences. While the hypothesis is supported, rejecting the null hypothesis, the substantial within-group variability suggests that factors beyond age significantly impact individual trust in aviation AI. This indicates a complex relationship between age and AI trust in specialized fields.

Our analysis of generational differences in AI trust reveals a complex picture. While generational identity accounts for approximately 3.6% of the variability in AI trust (eta-squared), more conservative estimates suggest an even smaller effect of 2.2% or as low as 0.6% (epsilon-squared and omega-squared). Some metrics' confidence intervals slightly dip below zero, indicating a potential phenomenon without definitive assertion. These findings align with our previous research (Halawi et al., 2024) on the interplay between perceived utility, trust, and AI adoption intentions. We extend this understanding by examining generational variations. McKnight et al. (2011) and Thatcher et al. (2011) emphasized human factors in fostering technology trust, which our results suggest manifest differently across age groups in aviation.

Rather than stark generational divides, we observe subtle differences within a complex attitudinal landscape. Generational affiliation plays a minor role in AI trust, with other factors like individual experiences, education, technological exposure, and personal values likely exerting stronger influences. Our study reveals nuanced generational variances in AI trust, highlighting the need for a multifaceted approach to understanding and fostering trust in AI technologies across different age demographics.

Conclusion

Understanding these generational differences is paramount when evaluating AI technologies' implementation and public acceptance in commercial aviation. The varying technological adoption rates and trust among different age groups may necessitate tailored approaches to training, communication, and system design to ensure widespread acceptance and optimal utilization of AI in aviation safety and operations. Despite the statistical significance observed, the small effect size implies that the practical implications of age-related differences in trust toward AI in aviation may be restricted. Other variables that should be addressed in this analysis might exert a more pronounced influence on trust levels. As indicated, allocating AI functions will require striking a balance that engenders trust and confidence in processes and outcomes. The layered issues for teaming between or among human operators and machines present profound developmental resolutions within the complex AI spectrum. Conversely, advances along the domain-specific trajectory could aid substantially in closing the trust gap.

Implications of the Study: Trust in AI is likely a dynamic construct that may change over time as AI technologies evolve and become more prevalent. Understanding age-independent factors in tech trust enables inclusive design and adoption strategies. This approach moves beyond stereotypes, recognizing the complex influences on our relationship with technology across generations.

Limitations of the Study and Future Directions: This research is exclusively centered on using AI within the aviation sector of the United States. The study specifically pertains to commercial airlines and explicitly excludes military aviation considerations. The analysis presupposes that aviation professionals will rationalize decision-making regarding adopting AI tools based on their perceived advantages and costs. The findings open up numerous avenues for further research and highlight the complexity of factors influencing trust in AI technologies. They underscore the importance of considering multiple variables, using diverse research methods, and focusing on equity and ethics in AI development and implementation. Longitudinal studies could provide valuable insights into how trust levels change and whether gender differences emerge or disappear as AI becomes more integrated into daily life.

References

Abbass, H. A. (2019). Social integration of artificial intelligence: Functions, automation allocation logic, and human-autonomy trust. *Cognitive Computation, 11*(2), 159–171. https://doi.org/10.1007/s12559-018-9619-0

Altman, Y., Baruch, Y., Zoghbi Manrique-de-Lara, P., & Viera Armas, M. M. (2020). Baby boomers at the cusp of their academic career: Storming ahead, Hanging on, or Calling it a day. *Studies in Higher Education* (Dorchester-on-Thames), *45*(7), 1335–1350. https://doi.org/10.1080/03075079.2019.1610864

Araujo, T., Helberger, N., Kruikemeier, S., & deVreese, C. (2020). In AI, we trust? Perceptions about automated decision-making by artificial intelligence. *AI & Society*, *35*, 611–623. https://doi.org/10.1007/s00146-019-00931-w

Borkovich, D. J. (2014). The influence of mobile technology culture: Blind trust, naivete, or skepticism. *Issues in Information Systems*, *15*(2), 399–410. https://doi.org/10.48009/2_iis_2014_399-410

Brooker, A. (2024, February 12). *What is AI, and why is it important for aviation?* Cirium. https://www.cirium.com/thoughtcloud/beyond-chatbots-how-ai-can-transform-aviation-and-challenges/

European Union Aviation Safety Agency (EASA). (2023). *Artificial intelligence roadmap 2.0: Human-centric approach to AI in aviation.* Version 2.0. www.easa.europa.eu/ai

Federal Aviation Administration (FAA). (2014). *Breakthroughs in major accident rates since 1946.* FAA Accident Investigation and Prevention.

Halawi, L., Miller, M., & Holley, S. (2024). Artificial intelligence in aviation: A path analysis. *Journal of Aviation/Aerospace Education and Research (JAAER)*, Special Issue Artificial Intelligence in Aviation, *33*(4), 1–9. https://commons.erau.edu/jaaer/vol33/iss4/10/

Hancock, P. A. (2023). Reacting and responding to rare, uncertain, and unprecedented events. *Ergonomics*, *66*(4), 454–478. https://doi.org/10.1080/00140139.2022.2095443

Harlianto, J., & Rudi Cheulie, L. (2020). Relationship analysis between key performance indicator with core competency for Baby Boomer, X, and Y Generation: Case Study: ABC Foundation. *Advances in Natural and Applied Sciences*, *14*(3), 46–53. https://www.aensiweb.net/AENSIWEB/anas/anas/2020/September/46-52(6).pdf

Holley, S., Miller, M., & Halawi, L. (2024). Artificial intelligence on the digital flight deck: A continuum with parallel trajectories. In *Proceedings of the Human Factors and Ergonomics Society Annual Meeting,* Phoenix, AZ. https://doi.org/10.1177/10711813241260403

Johnson, N. F. (2009). Generational differences in beliefs about technological expertise. *New Zealand Journal of Educational Studies, 44*(1), 31–45.

Johnson, M., & Vera, A. H. (2019). No AI is an Island: The case for teaming intelligence. *The AI Magazine*, *40*(1), 16–28. https://doi.org/10.1609/aimag.v40i1.2842

Kolmos, J. A. (2017). Human factors regarding age in single pilot transitions to technologically advanced aircraft. Collegiate Aviation Review, *35*(2). https://doi.org/10.22488/okstate.18.100477

Lazanyi, K. (2019). Generation Z and Y - are they different, when it comes to trust in robots? In *2019 IEEE 23rd International Conference on Intelligent Engineering Systems (INES)*, 000191–000194. https://doi.org/10.1109/INES46365.2019.9109508

Mcknight, D. H., Carter, M., Thatcher, J. B., & Clay, P. F. (2011). Trust in a specific technology: An investigation of its components and measures. *ACM Transactions on Management Information Systems*, *2*(2), 1–25. https://doi.org/10.1145/1985347.1985353

McKnight, D. H., Choudhury, V., & Kacmar, C. (2002). Developing and validating trust measures for e-commerce: An integrative typology. *Information Systems Research*, *13*(3), 334–359. https://doi.org/10.1287/isre.13.3.334.81

Miller, M., Holley, S., & Halawi, L. (2023, December 4–6). The evolution of AI on the commercial flight deck: Finding balance between efficiency and safety while maintaining the integrity of operator trust. In *14th AHFE International Conference on Human Factors in Design, Engineering, and Computing for All (AHFE 2023 Hawaii Edition)*, Waikiki Beach Marriott Resort & Spa, Honolulu, Hawaii, USA.

Miller, M., Holley, Sam, & Halawi, Leila (2024, September 9–13). POS137s1: Artificial intelligence on the digital flight deck: A continuum with parallel trajectories. In poster at the *ASPIRE - the HFES International Annual Meeting*, Phoenix, Arizona.

Nakagawa, K., & Yellowlees, P. (2020). Inter-generational effects of technology: Why millennial physicians may be less at risk for burnout than baby boomers. *Current Psychiatry Reports*, *22*(9), 45–. https://doi.org/10.1007/s11920-020-01171-2

Nørgaard, K., & Linden-Vørnle, M. (2021). Cyborgs, neuroweapons, and network command. *Scandinavian Journal of Military Studies*, *4*(1), 94–107. https://sjms.nu/articles/10.31374/sjms.86

Schubert, A., Löffler, C., Hagemann, D., & Sadus, K. (2023). How robust is the relationship between neural processing speed and cognitive abilities? *Psychophysiology*, *60*(2), e14165–. https://doi.org/10.1111/psyp.14165

Söllner, M. (2020). Same same but different? A two-foci perspective on trust in information systems. In *Proceedings of the 53rd Hawaii International Conference on System Sciences*, 5129–5138, University of Hawaii at Manoa, Hawaii. https://doi.org/10.24251/hicss.2020.630

Thatcher, J. B., McKnight, D. H., Baker, E. W., Arsal, R. E., & Roberts, N. H. (2011). The role of trust in postadoption IT exploration: An empirical examination of knowledge management systems. *IEEE Transactions on Engineering Management*, *58*(1), 56–70. https://doi.org/10.1109/TEM.2009.2028320

Thielsch, M. T., Meeßen, S. M., & Hertel, G. (2018). Trust and distrust in information systems at the workplace. *PeerJ*. https://peerj.com/articles/5483/

Warwick, G. (2024, June 17). Learning awareness. *Aviation Week and Space Technology, 186*(24), 3–31. https://aviationweek.com/aviation-week-space-technology/2024-06-17

Wing, D., Lacher, A., Ryan, W., Cotton, W., Stilwell, R., Maris, J., & Vajda, P. (2022). *Digital flight: A new cooperative operating mode to complement VFR and IFR*. Langley Research Center. https://ntrs.nasa.gov/api/citations/20220013225/downloads/NASA-TM-20220013225.pdf

Woodard, K., & Fairbrother, J. (2020). Cognitive loading during and after continuous task execution alters the effects of self-controlled knowledge of results. *Frontiers in Psychology, 11*. https://doi.org/10.3389/fpsyg.2020.01046

Yadav, R., & Chaudhari, S. (2024). Do leadership behaviour preferences differ among generations? A qualitative study of Gen X, Y, and Z in India. *Asian Business & Management*, *23*(2), 337–365. https://doi.org/10.1057/s41291-024-00266-5

11 Trust-based brand equity and generative AI

Hanna Górska-Warsewicz

Introduction

In recent years, there has been an intense growth in the importance of artificial intelligence (AI) in different areas, including branding and brand management issues (Ho & Chow, 2024; Kumar et al., 2019; Musaiqer & Hamdan, 2023). This is related to various aspects such as consumer loyalty to brands, choice factors, as well as consumer trust in brands. At the same time, there is an increase of the interest in the context of brand equity for various application areas (Dourado et al., 2020; Ha, 2020; Kazmi et al., 2021; Mega et al., 2018; Melina Dewi et al., 2018).

The growing importance of consumer trust towards companies and brands is analyzed in several dimensions. Reliability and intents are a couple of them (Delgado-Ballester & Munuera-Alemán, 2005; Delgado-Ballester, 2004; Luk & Yip, 2008). The role of brand reliability should be viewed through the prism of a brand's ability to deliver on its promises to customers. This guarantee defines the features and processes involved in its creation, production, marketing, delivery, and communication. Satisfaction of customer needs is dependent on the fulfillment of the promise (Delgado-Ballester & Munuera-Alemán, 2005).

Based on the above arguments, a research gap was identified regarding the combination of AI, including GenAI, brand equity, and brand trust; therefore, the aim of this research was to identify the importance of trust in the context of GenAI and brand equity based on a bibliometric analysis on brand equity and AI as well as brand equity and GenAI.

Three research questions were formulated:

- What is the state of development of the research area on AI and brand equity?
- What is the state of development of the research area on GenAI and brand equity?
- What is the importance of trust in both research areas, i.e., AI and brand equity as well as GenAI and brand equity?

DOI: 10.4324/9781003586937-14

Literature review

AI is transforming branding by improving customer engagement and brand management in several ways. This is linked, for example, to content personalization. By analyzing large datasets to forecast consumer behaviors and preferences, AI helps firms develop tailored marketing campaigns. As a result, marketing messages and products can be customized, increasing consumer happiness and engagement (Kumar et al., 2019; Musaiqer & Hamdan, 2023). With the use of AI-driven content strategies, customers may be hyper-targeted with offers, recommendations, and customized content. This strategy increases customer understanding and value, which in turn promotes brand loyalty (Singh & Kaunert, 2024). Content personalization is also achieved using interactive AI tools. Chatbots and virtual assistants are examples of branded AI systems that offer tailored interactions to improve client satisfaction. These solutions can provide recommendations and help in real time, streamlining and improving the client experience (Hassan et al., 2024).

An important area of application of AI in branding is content creation and storytelling. Quickly producing high-quality content is made easier with AI, and this is important for keeping up a strong online presence. AI-driven storytelling has the power to craft compelling stories that connect with viewers and improve brand loyalty and identification. This occurs through the use of AI in character creation for marketing and storytelling, as well as through the impact of generative AI-driven storytelling in marketing (Nimo et al., 2024).

The importance of AI also relates to customer experience. By providing individualized interactions and forecasting client pleasure, AI improves the customer experience. This results in stronger consumer-brand relationship and higher levels of brand loyalty. Studies cited in the literature show that brand experience, brand preference, and repurchase intention were impacted by AI marketing initiatives. Information, and accessibility (Ho & Chow, 2024). Al-based improved customer experience is achieved by providing personalized recommendations and automating repetitive tasks, which leads to a more efficient and satisfying customer experience. This not only improves customer satisfaction but also increases the likelihood of repeat purchases and brand endorsement (Alghizzawi et al., 2024; Armutcu et al., 2024).

AI is used in marketing through data analysis and customer profiling. AI algorithms can analyze vast amounts of data on consumer behavior, preferences, and interactions to create detailed customer profiles. This allows marketers to deliver highly targeted marketing messages that match individual interests (Hassan et al., 2024; Sahu & Mandal, 2024). Predictive analytics is also important. AI can predict customer behavior and preferences, allowing brands to anticipate needs and tailor their offers accordingly. This predictive ability helps create personalized experiences that increase customer satisfaction and loyalty (Singh & Kaunert, 2024; Venkateswaran et al., 2023).

AI is also used in shaping brand equity and identity. AI can redefine brand identity frameworks and optimize brand promotions. It helps in reformulating brand equity factors, ensuring that brands remain relevant and competitive in a rapidly evolving market (Kolla & Kumar, 2019). The issue of brand equity is widely analyzed in the literature in the context of defining its dimensions (Aaker, 1992; Keller & Lehmann, 2006) as well as determining the importance of brand equity for particular industries and economic sectors (Dourado et al., 2020; Górska-Warsewicz et al., 2021; Jacobsen, 2009; Konecnik & Gartner, 2007). Brand equity is comprised of four primary elements, namely brand loyalty, brand awareness, perceived quality, brand associations, and additional brand assets such as patents, trademarks, and distribution channel partnerships (Aaker, 1991). The influence of brand knowledge that distinguishes customer responses to marketing initiatives that work against the brand is the second idea, which refers to consumer brand equity. According to this perspective, brand knowledge consists of two components: brand image, which is created by strong, favorable, and distinctive associations with the brand and its characteristics, attitudes, and advantages (Keller, 1993). In this concept, AI helps develop brands by enhancing customer interaction, personalizing marketing techniques, and optimizing brand messaging (Musaiqer & Hamdan, 2023). In addition, AI transforms consumer journeys and marketing communications and content strategies, serving as a creative helper in content generation (Munyengeterwa et al., 2023)

The importance of trust in establishing and preserving brand equity is emphasized by trust-based brand equity. This idea is especially pertinent to peer-to-peer digital brands, where bilateral rating and review systems—as well as other trust mechanisms—are essential. In these circumstances, the development of brand equity is influenced by the combined effects of peer and brand trust (Ozbal et al., 2020). Consumers' prior experiences form the basis of brand trust, which is positively correlated with brand loyalty and, ultimately, builds brand equity (Delgado-Ballester & Munuera-Alemán, 2005). In addition, trust serves as a mediator between a variety of brand-related factors and brand equity, hence enhancing the relational aspect of brand equity (Ahmad & Thyagaraj, 2014; Delgado-Ballester & Munuera-Alemán, 2005).

Methodology

The aim of the study was to identify the importance of trust in the context of GenAI and brand equity based on a bibliometric analysis of brand equity and AI as well as brand equity and GenAI. The technique employed was bibliometric analysis, which is well liked in many academic and scientific studies (Bhukya et al., 2022; Donthu et al., 2021; Ellegaard, 2018; Moral-Muñoz et al., 2020; Rojas-Lamorena et al., 2022; Valenzuela-Fernandez et al., 2019).

A five-stage procedure (Zupic & Čater, 2015) was used, posing research questions in the first stage. In the second stage, a search was performed in the Scopus database, yielding 4,841 publications for issues related to AI and brand equity and 51 publications for issues related to GenAI and brand equity. The Scopus database was used for the search, adopting the reasoning from previous studies.

The search represents the status as of 15 September 2024. The following search paths were used (ALL("brand equity") AND ALL (AI) OR ALL ("artificial intelligence")) for AI and brand equity, and (ALL("brand equity") AND ALL ("Generative AI") OR ALL ("GenAI")) for GenAI and brand equity.

In the third step of the bibliometric analysis, quantitative indicators for the number of publications and citations and the average number of publications and citations per year were analyzed. The publications with the highest number of citations were also reported. The main authors, affiliation countries, document types, source titles, and keyword frequency were presented (Donthu et al., 2021). In the fourth stage, keyword co-occurrence mapping was also performed using VOSviewer (Nees Jan van Eck and Ludo Waltman, Leiden University, version 1.6.17) (van Eck & Waltman, 2010). In a fifth step, a discussion of the results was carried out by answering the research questions posed.

Results

Bibliometric analysis of AI and brand equity

The number of publications on AI and brand equity indexed in the Scopus database amounted to 4,841, with 2,342 publications from 2023 to 2024, accounting for 48.5% of all publications in this area. The number of publications of ten per year was exceeded in 2007 and 100 in 2017. Since that year, the number of publications has steadily increased, amounting to 150 publications in 2018, 213 in 2019, 322 in 2020, 506 in 2021, and 765 in 2022.

The first publication on AI and brand equity appeared in 1996 in the *Journal of Marketing Research.* The article, *Composite branding alliances: An investigation of extension and feedback effects,* analyzed various extensions of a composite brand in terms of their impact on consumer choice and preference (Park et al., 1996).

The total number of citations of AI and publications was 79,043, resulting in an h-index of 118. The differences between the number of citations in each year are significant. The annual number of citations exceeded 100 in 2009 and 1,000 in 2017. In subsequent years, the annual number of citations was 1,893 in 2018, 2,912 in 2019, 4,883 in 2020, 7,530 in 2021, 12,152 in 2022, and 19,463 in 2023. As of 15 September 2024, the number of citations was 25,078. The average annual number of citations was 3,040.

Two publications have more than 1,000 citations. The first publication is an article entitled *Marketing through Instagram influencers: the impact of number of followers and product divergence on brand attitude* published in the *International Journal of Advertising* in 2017. This publication has 1,134 citations. In two experimental studies, it was proven that influencers on Instagram with many followers are more liked (De Veirman et al., 2017). The second publication with a citation count of 1,043 is the article *The rise of motivational information systems: A review of gamification research* published in International Journal of Information Management in 2019. This chapter provides a comprehensive review of gamification research (819 studies in total) and analyses research models and empirical findings on gamification (Koivisto & Hamari, 2019).

Publications on AI and brand equity are classified into the following subject areas: business, management, and accounting – 2,808 publications, computer science – 1,477, and social sciences – 1,298. Other subject areas to economics, econometrics, and finance – 890 studies, engineering – 590, decision sciences – 481, psychology – 353, environmental science – 337, mathematics – 330, arts and humanities – 235, and energy – 227. Other subject areas were represented by fewer than 100 publications, including medicine, agricultural and biological sciences, multidisciplinary, materials science, and others.

In terms of document type, publications in the field of AI and brand equity were primarily articles (3,531), conference papers (479), book chapters (430), books (201), reviews (187), editorials (10), notes (6), and conference reviews (5). Accordingly, studies were published in journals (3,740 studies), books (509), book series (301), conference proceedings (301), and trade journals (2). Most publications appeared in *Sustainability* (152), *Journal of Retailing and Consumer Services* (120), *Lecture Notes in Computer Science Including Subseries Lecture Notes in Artificial Intelligence and Lecture Notes in Bioinformatics* (116), *Journal of Business Research* (80), *Frontiers in Psychology* (59), and *Studies in Computational Intelligence* (52). Further positions were taken by, among others: *Journal of Product and Brand Management* (38), *Technological Forecasting and Social Change* (38), *Psychology and Marketing* (36), *International Journal of Hospitality Management* (35), *Asia Pacific Journal of Marketing and Logistics* (32), *Asia Pacific Journal of Marketing and Logistics* (32), and *Heliyon* (30). Other journals have dealt with consumer behavior, including the *Journal of Consumer Behavior* (28) and *International Journal of Consumer Studies* (26).

Authors of AI and brand equity publications indexed in the Scopus database were primarily affiliated with countries such as China (782), the United States (732), India (664), the United Kingdom (428), Malaysia (330), South Korea (247), Australia (231), Taiwan (229), and Spain (207). To a lesser extent, the authors of the publications were from Italy (187), Germany (168), Indonesia (159), United Arab Emirates (149), France (138), Jordan (137), Saudi Arabia (134), Turkey (131), Portugal (121), Pakistan (117), and Canada (107). The following countries were represented in 50–100 publications: VietNam, Hong Kong, Iran, Thailand, the Netherlands, Poland, Greece, South Africa, Bangladesh, Japan, and Finland.

The authors with the highest number of publications were M.T. Alshurideh (33 studies), S.I.S. Al-Hawary (31), H.C. Wu (24), P. Foroudi (22), M. Alshurideh (21), and H.M Alzoubi (20). Fewer publications were authored by the following researchers: N.P. Rana (19), C.C. Cheng (17), A.A.S. Mohammad (17),

M.T. Nuseir (17), Y.K. Dwivedi (16), S.M.C. Loureiro (16), B. Al Kurdi (15), H. Han (14), S. Kumar (14), F.M.F. Aldaihani (13), A. Behl (13), S. Chatterjee (13), D. Vrontis (13), N.S. Alajarmeh (12), M. Alzyoud (12), R. Chaudhuri (12), N. Al-shanableh (11), J. Hwang (11), W.M. Lim (11), and S. Urabi (11). Most of the publications were published in English (4,801), but the Scopus database also indexed publications in Spanish (25), Chinese (14), French (6), Korean (5), Portuguese (5), Italian (4), Japanese (2), and Turkish (2).

The following keywords dominated in terms of frequency: social media (371), marketing (320), AI (295), sales (262), and commerce (205). Keywords such as social networking (197), consumer behavior (164), decision-making (161), purchase intention (150), electronic commerce (147), consumption behavior (141), customer satisfaction (130), trust (128), human (123), sustainability (123), virtual reality (110), perception (103), brand equity (101), and machine learning (101) were frequently used. Keywords appearing in 75–100 publications are article, sustainable development, COVID-19, satisfaction, tourism, service quality, innovation, behavioral research, surveys, digital marketing, and customer experience. A smaller number of publications (50–75) included keywords related to brand (brand image, brand loyalty), marketing (social media marketing and perceived value), customers (customer engagement and customer loyalty), sales (retailing and purchasing), AI (augmented reality, data mining, and human computer interaction), methodological methods (literature review, bibliometric analysis, systematic literature review, sentiment analysis, and structural equation modelling) as well as economic and social effects, competition, and corporate social responsibility.

The analysis of co-occurrence of keywords made it possible to distinguish ten clusters based on the fractional counting and the association strength method with a minimum number of occurrences of a keyword of six (Figure 11.1):

- Cluster No. 1 (161 items) with examples of keywords such as experience, behavioral intensions, environmental protection, preference behavior, product innovation, perception, risk perception, and satisfaction.
- Cluster No. 2 (148) with examples of keywords related to brands (brand personality, brand perception), consumers (consumer experience, perception, purchase, experience, intensions), information (use, systems, sources), and mobile aspects (apps, commerce, communications) and gamification, virtual reality, worlds, and communication.
- Cluster No. 3 (136) with examples of keywords related to advanced technology, AI technology, big data, brands (recognition and value), customer (satisfaction, preferences, segmentations, reviews), deep learning, digital technologies, social media machine learning, marketing research and strategies, product design and development.
- Cluster No. 4 (106) with examples of keywords related to business (model, strategy, environment, value, and performance), consumer (behavior,

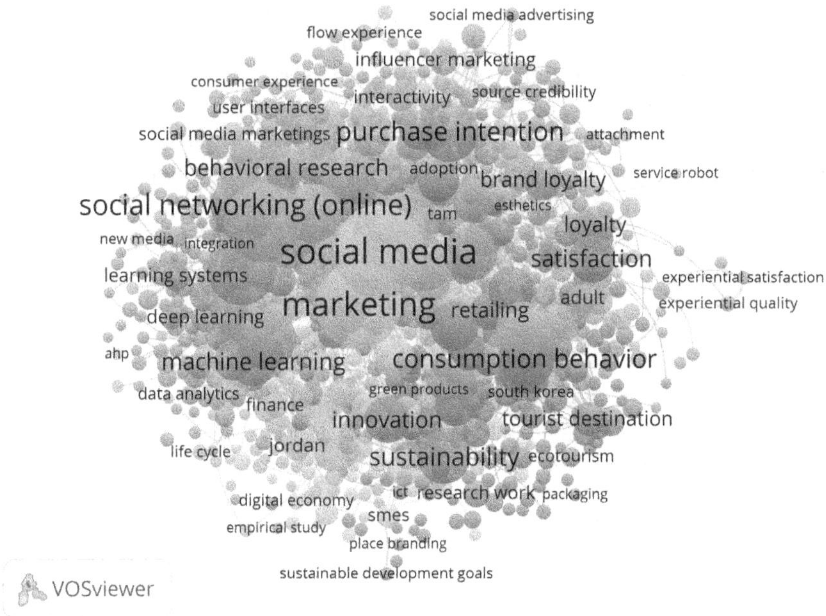

Figure 11.1 Co-occurrence of keywords in publications related to AI and brand equity

Source: VOSviewer based on Scopus database, 15.09.2024

psychology, response, engagement), digital issues (communication, economy, marketing, technology, and transformation), marketing (communication, mix, performance, and tools), and social networks and social media.

- Cluster No. 5 (100 items) with examples of keywords related to brand (attachment, attitude, awareness, communication, love, loyalty, identity, equity, experience, satisfaction, trust, strategy, knowledge, image, identification, preference, and identification), corporate (branding, reputation, social responsibility), digital platforms, mobile apps, social media (marketing, influences, and communities), and trust.
- Cluster No. 6 (89 items) with examples of keywords related to competition and competitive advantage, economic aspects, emerging economies, information management, industry 4.0, and manufacture.
- Cluster No. 7 (74 items) with such keywords like emotions and emotional response, neuroscience, quality, and quality management, transparency, and crowdsourcing.
- Cluster No. 8 (61 items) with examples of keywords related to brand (authenticity, credibility, engagement, and name), customer (engagement, experience, journey, and value), product (evaluation and attributes), social media advertising, and IoT.

- Cluster No. 9 (55 items) with examples of keywords related to AI, storytelling, trustworthiness, customer (satisfaction, trust, service, loyalty), customization, chatbots, luxury and luxury brands, and service (robots, quality, value, and experience).
- Cluster No. 10 (21 items) with examples of keywords related to brand management, personal branding, creativity, new product development, cultural heritage, and marketing management.

Bibliometric analysis of GenAI and brand equity

The number of publications indexed in the Scopus database on GenAI and brand equity was 51. These are publications from 2023 and 2024, with four publications in 2023 and 47 publications in 2024. The number of citations was 139 for 18 publications, resulting in an h-index of 5.

The publication indexed in the Scopus database having the most citations was the article *How to Bell the Cat? A Theoretical Review of Generative Artificial Intelligence towards Digital Disruption in All Walks of Life.* The article examines the applications of GenAI in various sectors for developing effective business strategies. New customer propositions relate to the creation of customer value and experience and propositions in information technology and digital communications. Business management solutions for enhancing organizational effectiveness were proposed (Mondal et al., 2023).

The second publication in terms of number of citations is *The Caring Machine: Feeling AI for Customer Care*, published in *Journal of Marketing* in 2024. The paper outlines the potential of GenAI for customer care, which connects the fields of marketing and IT. The customer care pathway using AI from emotion recognition, including response, and establishing an emotional connection was analyzed. Marketing principles for implementing the use of GenAI in customer service were proposed (Huang & Rust, 2024). The third article entitled *To ChatGPT, or not to ChatGPT: Navigating the paradoxes of generative AI in the advertising industry,* addresses the use of GenAI in advertising. The article examines text-to-text GenAI in relation to ChatGPT in the advertising industry and presents a psychological paradox related to professional identity (Osadchaya et al., 2024).

Publications indexed in the Scopus database were classified into subject areas such as business, management, and accounting (30 studies), social sciences (20), computer science (13) and economics, econometrics, and finance (13). Other subject areas included fewer than ten publications, including psychology (four studies), engineering (four), mathematics (three), multidisciplinary (two), energy (two), arts and humanities (two), environmental sciences (one), decision sciences (one), as well as agricultural and biological sciences (one).

In terms of document types, publications indexed in the Scopus database are primarily articles (32), books (7), book chapters (5), reviews (4), and conference papers (3). Studies were published in journals (36 studies), books (12), book series (2), and conference proceedings (2). Among the source titles were *Current Issues in Tourism* (2 studies), *Journal of Consumer Behaviour* (2), *Journal of Ecohumanism* (2), *Journal of Retailing and Consumer Services* (2), and *Leveraging Chatgpt and Artificial Intelligence for Effective Customer Engagement* (2). Other source titles appeared one at a time, including *Business Strategy and Development, Communications in Computer and Information Science, Communications of the Association for Information Systems, Consumer Behaviour and Digital Transformation, International Journal of Market Research, Journal of Advertising Research, International Journal of Gastronomy and Food Science,* and *Marketing Education Review.*

Publications affiliated with the Scopus database were represented by researchers from countries such as the United States (15 studies), India (6), and South Korea (4). Authors of a smaller number of publications came from other countries, including China, Hong Kong, and Malaysia (3 publications each), and Greece, Indonesia, Morocco, the Netherlands, Portugal, Spain, and Taiwan (2 publications each). Two researchers: S.W. Litvin and K.L. Siau authored two publications. All studies were published in English.

The main keywords for publications indexed in the Scopus database were AI (13), ChatGPT (7), generative AI (5), and generative AI (4). Keywords such as algorithm, brand personality, chatbots, marketing, and social media appeared in three publications. In contrast, keywords such as branding, brand equity, brand image, brand love, brand trust, brand voice, brand self-congruity, brand anthropomorphism, customer engagement, customer involvement, customer relationship marketing, and customer satisfaction appeared in single studies.

Analysis of co-occurrence of keywords using fractional counting and the association strength method with a minimum number of occurrences of a keyword of two allowed four clusters to be selected (Figure 11.2):

- Cluster No. 1 (8 items) with keywords like algorithm, algorithms, algorithm aversion, AI, human, live streaming, and social media.
- Cluster No. 2 (7 items) with AI, chatbots, ChatGPT, commerce, and generative AI.
- Cluster No. 3 (3 items) with keywords like brand personality, e-wom, and service quality.
- Cluster No. 4 (2 items) with keywords like consumption behavior and generative AI.

Importance of trust for AI and GenAI research areas, including brand equity

In the area of GenAI and brand equity, trust as a keyword appeared in 128 publications and brand trust in 30 publications. Customer trust was analyzed in the

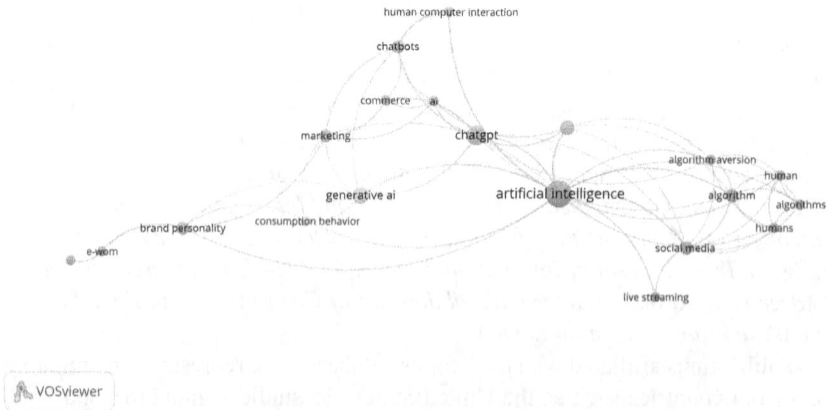

Figure 11.2 Co-occurrence of keywords in publications related to GenAI and brand equity

Source: VOSviewer based on Scopus database, 15.09.2024

context of customer engagement, customer retention, customer citizenship behavior, customer satisfaction, and perceived economic benefits. Brand trust, on the other hand, was analyzed in relation to brand loyalty. The creation of relationships based on trust, the post-regulation of trust in the context of risk, and the role of trust in brand image building were also pointed out. Trust in AI and brand equity were also analyzed in relation to brand experience, brand attachment in social media. In the context of AI, the role of brand communities in social media marketing was also important. For example, the studies analyzed factors responsible for creating brand trust and brand distrust among consumers (Mal et al., 2018), the influence of sustainable marketing practices on consumers' attitudinal and behavioral brand loyalty (Khandai et al., 2023), and the impact of customer engagement and service evaluation on brand trust and customer behavioral intention (Chen et al., 2022).

In the analysis of the co-occurrence of keywords across the ten identified clusters, the issue of trust was present in two clusters. In Cluster No. 5, trust was analyzed both in general terms and in relation to the brand, alongside other brand attributes concerning attachment, attitude, awareness, communication, love, loyalty, identity, equity, experience, satisfaction, trust, strategy, knowledge, image, identification, preference, and identification. On the other hand, in Cluster No. 9, trustworthiness and customer trust were analyzed alongside customer satisfaction, service, and loyalty.

In the area of GenAI and brand equity, trust as a keyword appeared in one article, while the topic of trust was addressed in six publications. For example, AI avatars were analyzed in relation to communication, psychology, and social psychology, pointing to their ease of customization and personalization (Liu & Siau, 2023). The impact of the metaverse on consumer behavior was also investigated with an integrated behavioral model, consumer perceptions of metaverse opportunities, and the role of trust and flow experiences in metaverse shopping platforms (Zhong & Hamouda, 2024). Decoding the impact of social media on consumer choices in the context of brand equity (Springer Proceedings, 2024) and the mediating effect of brand trust on the relationship between brand personality, brand self-congruence, and brand love and E-WOM (Juhaidi et al., 2024) have also been studied.

Conclusions

This chapter has made it possible to identify the degree of development of the research space on AI, GenAI, and brand equity. This involved answering the first two research questions. The first research area on AI and brand equity is represented by more than 4,800 publications indexed in the Scopus database with a significant number of citations of more than 79,000. In contrast, the second research area on GenAI and brand equity has 51 publications with less than 140 citations. This implies a significant predominance of publications in the first area, while it should be noted that the second research area combining GenAI and brand equity is new, innovative, and characterized by interdisciplinarity.

The issue of trust regarding AI, GenAI, and brand equity is an important emerging issue in both areas in the context of the consumer-brand relationship. This has been considered for management decisions, building consumer loyalty, and shaping long-term stakeholder relationships. Modern opportunities in social media marketing, influencer marketing, and shaping brand communities were considered. Furthermore, the issue of consumer trust in a brand is not only analyzed in relation to AI or brand equity but also in the spectrum of a wide range of issues, including brand attachment, loyalty, identity, attitude, awareness, communication, love, experience, satisfaction, and brand strategy, knowledge, image, identification, preference, and identification.

Identifying brand trust relationships in the context of other brand attributes, AI and GenAI, should be pointed as directions for future research. It should also be important to analyze the relationships between AI, GenAI, and branding in different economic sectors and for different audiences.

References

Aaker, D. (1991). *Managing Brand Equity: Capitalizing on the Value of a Brand Name*. New York: Free Press; Toronto: Maxwell Macmillan Canada; New York: Maxwell Macmillan International, ©1991.

Aaker, D. (1992). The value of brand equity. *Journal of Business Strategy*, *13*(4), 27–32. https://doi.org/10.1108/eb039503

Ahmad, A., & Thyagaraj, K. S. (2014). Evaluating the role of consumer brand relationships in brand equity formation. *International Journal of Applied Business and Economic Research*, *12*(3), 603–614.

Alghizzawi, M., Ahmed, E., Ezmigna, I., Ezmigna, A. A. R., & Omeish, F. (2024). The relationship between artificial intelligence and digital marketing in business companies. *Studies in Systems, Decision and Control*, *525*, 885–895. https://doi.org/10.1007/978-3-031-54383-8_67

Armutcu, B., Tan, A., Ho, S. P. S., Chow, M. Y. C., & Gleason, K. C. (2024). The effect of bank artificial intelligence on consumer purchase intentions. *Kybernetes*, *ahead-of-print* (ahead-of-print). https://doi.org/10.1108/K-01-2024-0145

Bhukya, R., Paul, J., Kastanakis, M., & Robinson, S. (2022). Forty years of European management journal: A bibliometric overview. *European Management Journal*, *40*(1), 10–28. https://doi.org/10.1016/j.emj.2021.04.001

Chen, X., Wang, Y., Lyu, X., & Zhang, J. (2022). The impact of hotel customer engagement and service evaluation on customer behavior intention: The mediating effect of brand trust. *Frontiers in Psychology*, *13*, 852336. https://doi.org/10.3389/fpsyg.2022.852336

De Veirman, M., Cauberghe, V., & Hudders, L. (2017). Marketing through Instagram influencers: The impact of number of followers and product divergence on brand attitude. *International Journal of Advertising*, *36*(5), 798–828. https://doi.org/10.1080/02650487.2017.1348035

Delgado-Ballester, E. (2004). Applicability of a brand trust scale across product categories. *European Journal of Marketing*, *38*(5/6), 573–592. https://doi.org/10.1108/03090560410529222

Delgado-Ballester, E., & Munuera-Alemán, J. L. (2005). Does brand trust matter to brand equity? *Journal of Product and Brand Management*, *14*(3), 187–196. https://doi.org/10.1108/10610420510601058

Donthu, N., Kumar, S., Mukherjee, D., Pandey, N., & Lim, W. M. (2021). How to conduct a bibliometric analysis: An overview and guidelines. *Journal of Business Research*, *133*(March), 285–296. https://doi.org/10.1016/j.jbusres.2021.04.070

Dourado, P., Palma, L., Maria Pinto da Cunha Brandão, A., & Palma Martos, L. (2020). See now, buy now model: A passport to fashion brand equity. *Scientific Annals of Economics and Business*, *December*. https://doi.org/10.47743/saeb-2020-0028

Ellegaard, O. (2018). The application of bibliometric analysis: Disciplinary and user aspects. *Scientometrics*, *116*(1), 181–202. https://doi.org/10.1007/s11192-018-2765-z

Górska-Warsewicz, H., Dębski, M., Fabuš, M., & Kováč, M. (2021). Green brand equity—empirical experience from a systematic literature review. *Sustainability 2021*, *13*(20), 11130. https://doi.org/10.3390/SU132011130

Ha, M. T. (2020). Investigating green brand equity and its driving forces. *Management Science Letters*, *10*(10), 2379–2384. https://doi.org/10.5267/j.msl.2020.2.026

Hassan, A., Mohammed, F. A., & Seyadi, A. Y. (2024). Artificial intelligence applications for marketing. *Studies in Systems, Decision and Control*, *528*, 607–618. https://doi.org/10.1007/978-3-031-56586-1_43

Ho, S. P. S., & Chow, M. Y. C. (2024). The role of artificial intelligence in consumers' brand preference for retail banks in Hong Kong. *Journal of Financial Services Marketing*, *29*(2), 292–305. https://doi.org/10.1057/s41264-022-00207-3

Huang, M. H., & Rust, R. T. (2024). The caring machine: Feeling AI for customer care. *Journal of Marketing*, *88*(5), 1–23. https://doi.org/10.1177/00222429231224748

Jacobsen, B. P. (2009). Investor-based place brand equity: A theoretical framework. *Journal of Place Management and Development*, *2*(1), 70–84. https://doi.org/10.1108/17538330910946029

Juhaidi, A., Syaifuddin, Salamah, Ma'ruf, H., Yuseran, M., Shapiah, Janah, R., Mudhiah, Saputri, R. A., Muharramah, N., Fuady, M. N., Ramadan, W., & Lisdariani, R. (2024). The effect of brand personality, brand-self congruity, and brand love on E-WOM in Islamic higher education in Indonesia: A mediating effect of brand trust. *Social Sciences and Humanities Open*, *10*, 100955. https://doi.org/10.1016/j.ssaho.2024.100955

Kazmi, S. H. A., Shahbaz, M. S., Mubarik, M. S., & Ahmed, J. (2021). Switching behaviors toward green brands: Evidence from emerging economy. *Environment, Development and Sustainability*, 0123456789. https://doi.org/10.1007/s10668-020-01116-y

Keller, K. L. (1993). Conceptualizing, measuring, and managing customer-based brand equity. *Journal of Marketing*, *57*(1), 1. https://doi.org/10.2307/1252054

Keller, K. L., & Lehmann, D. R. (2006). Brands and branding: Research findings and future priorities. *Marketing Science*, *25*(6), 740–759. https://doi.org/10.1287/mksc.1050.0153

Khandai, S., Mathew, J., Yadav, R., Kataria, S., & Kohli, H. (2023). Ensuring brand loyalty for firms practising sustainable marketing: A roadmap. *Society and Business Review*, *18*(2), 219–243. https://doi.org/10.1108/SBR-10-2021-0189

Koivisto, J., & Hamari, J. (2019). The rise of motivational information systems: A review of gamification research. *International Journal of Information Management*, *45*, 191–210. https://doi.org/10.1016/J.IJINFOMGT.2018.10.013

Kolla, N., & Kumar, M. G. (2019). Meta-synthesis on artificial intelligence (AI): Imperatives for branding. *International Journal of Recent Technology and Engineering*, *8*(3), 2251–2255. https://doi.org/10.35940/ijrte.B3268.098319

Konecnik, M., & Gartner, W. (2007). Customer-based brand equity for a destination. *Annals of Tourism Research*, *34*(2), 400–421. https://doi.org/10.1016/j.annals.2006.10.005

Kumar, V., Rajan, B., Venkatesan, R., & Lecinski, J. (2019). Understanding the role of artificial intelligence in personalized engagement marketing. *California Management Review*, *61*(4), 135–155. https://doi.org/10.1177/0008125619859317

Liu, Y., & Siau, K. L. (2023). Human-AI interaction and AI avatars. In *Lecture Notes in Computer Science (Including Subseries Lecture Notes in Artificial Intelligence and Lecture Notes in Bioinformatics)*, vol. 14059, 120–130. https://doi.org/10.1007/978-3-031-48057-7_8

Luk, S. T. K., & Yip, L. S. C. (2008). The moderator effect of monetary sales promotion on the relationship between brand trust and purchase behaviour. *Journal of Brand Management*, *15*(6), 452. https://doi.org/10.1057/bm.2008.12

Mal, C. I., Davies, G., & Diers-Lawson, A. (2018). Through the looking glass: The factors that influence consumer trust and distrust in brands. *VPsychology and Marketing*, *35*(12), 936–947. https://doi.org/10.1002/mar.21146

Mega, M., Dewi, D., Anandya, C., & Rahardja, H. (2018). Building local fashion brand equity for young consumers. *Advances in Social Science, Education and Humanities Research (ASSEHR) 15th International Symposium on Management*, *186*, 166–169. https://doi.org/10.2991/INSYMA-18.2018.41

Melina Dewi, M., Anandya, D., & Rahardja Honantha, C. (2018). Building local fashion brand equity for young consumers. In 186*(Insyma)*. https://doi.org/10.2991/insyma-18.2018.41

Mondal, S., Das, S., & Vrana, V. G. (2023). How to bell the cat? A theoretical review of generative artificial intelligence towards digital disruption in all walks of life. *Technologies 2023*, *11*(2), 44. https://doi.org/10.3390/TECHNOLOGIES11020044

Moral-Muñoz, J. A., Herrera-Viedma, E., Santisteban-Espejo, A., & Cobo, M. J. (2020). Software tools for conducting bibliometric analysis in science: An up-to-date review. *Profesional de La Información, 29*(1), 1699–2407. https://doi.org/10.3145/EPI.2020.ENE.03

Munyengeterwa, T. S., Richards, M. B., & Eaton, J. B. (2023). Artificial intelligence trends and perceptions: Content strategy and the customer journey. In *Multidisciplinary Approaches in AI, Creativity, Innovation, and Green Collaboration*, 88–120. https://doi.org/10.4018/978-1-6684-6366-6.CH006

Musaiqer, H. H., & Hamdan, A. (2023). The role of artificial intelligence in brand building: A review. *Contributions to Management Science, Part F1640*, 307–318. https://doi.org/10.1007/978-981-99-6101-6_22

Nimo, A., Ravishankar, K., & Rajagopal, N. K. (2024). AI for character creation and storytelling in marketing. In A. Nair, M. Sridhar, A. Mittal, & A. Wasim (Eds.), *Balancing Automation and Human Interaction in Modern Marketing* (pp. 39–58). IGI Global. https://doi.org/10.4018/979-8-3693-2276-5.CH003

Osadchaya, E., Marder, B., Yule, J. A., Yau, A., Lavertu, L., Stylos, N., Oliver, S., Angell, R., Regt, A. de, Gao, L., Qi, K., Zhang, W. Z., Zhang, Y., Li, J., & AlRabiah, S. (2024). To ChatGPT, or not to ChatGPT: Navigating the paradoxes of generative AI in the advertising industry. *Business Horizons, 67*(5), 571–581. https://doi.org/10.1016/J.BUSHOR.2024.05.002

Ozbal, O., Duman, T., & Topaloglu, O. (2020). A trust-based peer-to-peer digital brand equity (P2P-DBE) model. *Journal of Marketing Theory and Practice, 28*(4), 497–520. https://doi.org/10.1080/10696679.2020.1794901

Park, C. W., Jun, S. Y., & Shocker, A. D. (1996). Composite branding alliances: An investigation of extension and feedback effects. *Journal of Marketing Research, 33*(4), 453–466. https://doi.org/10.2307/3152216

Rojas-Lamorena, Á. J., Del Barrio-García, S., & Alcántara-Pilar, J. M. (2022). A review of three decades of academic research on brand equity: A bibliometric approach using co-word analysis and bibliographic coupling. *Journal of Business Research, 139*, 1067–1083. https://doi.org/10.1016/j.jbusres.2021.10.025

Sahu, P., & Mandal, P. (2024). Unleashing the power of customer personalization in the digital age with artificial intelligence. In *Improving Service Quality and Customer Engagement with Marketing Intelligence*, 97–113. https://doi.org/10.4018/979-8-3693-6813-8.CH004

Singh, B., & Kaunert, C. (2024). Future of digital marketing: Hyper-personalized customer dynamic experience with AI-based predictive models. In *Revolutionizing the AI-Digital Landscape: A Guide to Sustainable Emerging Technologies for Marketing Professionals*, 189–208. https://www.taylorfrancis.com/chapters/edit/10.4324/9781032688305-14/future-digital-marketing-hyper-personalized-customer-dynamic-experience-ai-based-predictive-models-bhupinder-singh-christian-kaunert?context=ubx&refId=a88ed3fe-45d1-422a-991d-935795193e07

Springer proceedings. (2024). In *5th International Conference on Digital Marketing and eCommerce Conference, DMEC 2024. Springer Proceedings in Business and Economics*, Barcelona.

Valenzuela-Fernandez, L., Merigó, J. M., Lichtenthal, J. D., & Nicolas, C. (2019). A bibliometric analysis of the first 25 years of the journal of business-to-business marketing. *Journal of Business-to-Business Marketing, 26*(1), 75–94. https://doi.org/10.1080/1051712X.2019.1565142

van Eck, N. J., & Waltman, L. (2010). Software survey: VOSviewer, a computer program for bibliometric mapping. *Scientometrics*, *84*(2), 523–538. https://doi.org/10.1007/S11192-009-0146-3

Venkateswaran, P. S., Lishmah Dominic, M., Agarwal, S., Oberai, H., Anand, I., & Rajest, S. S. (2023). The role of artificial intelligence (AI) in enhancing marketing and customer loyalty. In *Data-Driven Intelligent Business Sustainability*, 32–47. https://doi.org/10.4018/979-8-3693-0049-7.CH003

Zhong, Z., & Hamouda, M. (2024). The impact of immersive and flow experiences on consumer participation in hyper-connected shopping platforms: A metaverse perspective. *Journal of Consumer Behaviour*. https://doi.org/10.1002/cb.2377

Zupic, I., & Čater, T. (2015). Bibliometric methods in management and organization. *Organizational Research Methods*, *18*(3), 429–472. https://doi.org/10.1177/1094428114562629

12 Healing and harmful powers of data

Trust in generative ai within healthcare

Angelica Marotta

Introduction: the AI revolution in healthcare

In the fast-paced world of healthcare, generative Artificial Intelligence (AI) is rapidly becoming a transformative tool that can reshape how patient care is delivered. In recent years, the market size expanded from $6.7 billion in 2020 to $22.4 billion in 2023, representing a staggering 233% increase (AIPRM, 2024). This trajectory is set to continue, with projections indicating that the global AI in healthcare market will reach $208.2 billion by 2030, a 524% growth from 2024 (AIPRM, 2024). The United States is poised to lead this revolution, with forecasts suggesting it will generate $102.2 billion in revenue through AI in the health sector by 2030 (AIPRM, 2024). This rapid expansion underscores the potential of AI in addressing some of the most challenging healthcare issues, such as high costs, lengthy drug development timelines, and provider burnout. Generative AI is finding applications across various healthcare domains, each at different stages of maturity. For instance, in drug development, approximately 70 drugs developed with assistance from generative AI were in clinical trials as of December 2023, potentially accelerating the drug discovery process (U.S. Government Accountability Office, 2024). Additionally, in clinical practice, AI is being employed to synthesize patient notes, a development that promises to alleviate administrative burdens and mitigate burnout among medical professionals (Shaheen, 2021). This technology is also transforming clinical trials by generating digital patient models from historical data, thereby enhancing the statistical power and reliability of smaller studies (Schott et al., 2015). Another example is diagnostic imaging; AI algorithms can significantly improve the quality of medical scans by enhancing diagnostic accuracy (Sanford et al., 2020). Furthermore, a study published in the New England Journal of Medicine indicated that generative AI has improved patient outcomes by up to 45% in clinical trials, particularly in treating chronic diseases through individualized management plans (Riddleberger, 2024).

However, while the potential benefits are substantial, implementing generative AI in healthcare also presents notable challenges. For example, reliability concerns arise from the potential for AI models to produce plausible but inaccurate

DOI: 10.4324/9781003586937-15

outputs, which could be dangerous in medical settings (Davenport & Kalakota, 2019). Using sensitive health data also raises concerns about patient privacy and compliance with healthcare data protection laws (Marotta & Madnick, 2022). AI models require large volumes of high-quality data, which can be challenging to obtain in healthcare due to fragmented systems and inconsistent data formats. If training data are not representative, AI outputs may be biased and underperform for underrepresented groups (Sperling, 2019). Public trust remains a significant hurdle, with a survey finding that 60% of patients would feel uncomfortable if their healthcare provider relied on AI, indicating a need for public education and trust-building (AIPRM, 2024). In addition, the rapid advancement of AI technology, particularly in healthcare, has prompted governments worldwide to grapple with the challenge of regulation (Marotta et al., 2023). The European Union (EU) has been at the forefront of these efforts with its Artificial Intelligence Act, aiming to establish the first comprehensive AI regulation across the EU (European Union, 2024). This pioneering legislation could have far-reaching implications for developing and deploying AI in healthcare globally.

This overview sets the stage for a comprehensive analysis of this powerful technology in the medical field. In particular, this chapter examines the dual impact of generative AI in healthcare, exploring its benefits and challenges. Building on this dichotomy, the research question for this study investigates the specific conditions that determine whether AI enhances or hinders healthcare development. To achieve this goal, this chapter is organized as follows: First, it provides an overview of the current academic state of the art regarding AI in healthcare, highlighting its transformative potential. Next, it delves into a comprehensive analysis of AI implementation's benefits and limitations, using the Hospital Clínic de Barcelona (HCB) case as a real-world example. In the final section, this chapter identifies critical areas for improvement and offers practical strategies to optimize AI's impact. These recommendations aim to help healthcare institutions navigate the complexities of AI adoption while minimizing potential drawbacks.

Literature review: mapping the terrain of generative AI in medicine

The integration of generative AI in healthcare has been a subject of intense research and debate in recent years, marking a significant shift in how medical professionals approach patient care, diagnosis, and treatment planning (Bates et al., 2020; Rong et al., 2020; Shah, 2019). However, its implementation is not without controversy and complexities, which have been investigated by several scholars (Hu et al., 2020). This literature review examined recent research on the applications, benefits, limitations, and ethical considerations surrounding generative AI in healthcare contexts. By synthesizing findings from diverse studies, this review aims to explore studies featuring cutting-edge applications while delving into critical discussions about challenging aspects of AI.

Applications and benefits

As AI technology continues to evolve, its applications in healthcare are becoming increasingly diverse and impactful. Numerous studies have documented its applications (Schwendicke & Krois, 2021; Shaheen, 2021). These research findings consistently highlight the technology's capacity to process vast amounts of medical data and generate novel insights. While acknowledging the transformative potential, researchers also emphasize the need for rigorous validation and ethical considerations in AI deployment (Bleher & Braun, 2022; Lysaght et al., 2019). This growing body of evidence underscores generative AI's significant contributions to healthcare, suggesting a paradigm shift in medical practices and scientific approaches.

- **Personalized Medicine and Treatment**: Recent studies highlighted generative AI's potential to enable more personalized approaches to medicine. As noted by Reddy (2024), generative AI has emerged as a transformative tool in healthcare, with the capacity to revolutionize clinical decision-making and improve health outcomes. This capability allows for developing individualized treatment programs based on patients' unique genetic profiles and medical histories, potentially improving prognoses and transforming care delivery. Expanding on this concept, Ghebrehiwet et al. (2024) conducted a systematic review that underscores the revolutionary impact of generative AI on precision medicine. Their study emphasizes how generative AI, particularly Deep Generative Models, contributes to advancing personalized healthcare by generating high-quality synthetic data, enhancing diagnostic accuracy, and facilitating the development of customized treatment plans. This review also highlighted the exponential growth of publications on generative AI in personalized medicine, with a 25-fold increase from 2014 to 2023, indicating the rapidly growing interest and potential in this field (Ghebrehiwet et al., 2024).
- **Diagnostic Assistance**: Multiple researchers point to AI's growing role in medical diagnostics (De Fauw et al., 2018; Mun et al., 2021; Sampat et al., 2005). For example, studies have shown that, in some cases, AI programs can outperform dermatologists in diagnosing skin cancer with greater speed and efficiency (Anderson et al., 2023; Salinas et al., 2024). In radiology, several investigations found that AI-based diagnostic algorithms can assist in breast cancer detection from mammograms, serving as a possible "second opinion" for radiologists (Bargalló et al., 2014; Garbin & Marques, 2022; iSanidad, 2023; Mun et al., 2021; Sampat et al., 2005). For instance, a large-scale study involving over 25,000 mammograms showed that an AI system reduced false positives by 5.7% and false negatives by 9.4% when used as a second reader alongside radiologists (McKinney et al., 2020). However, studies on this topic generally concur that AI systems are intended to complement rather than replace expert radiologists; AI-assisted readings tend to demonstrate either reduced or, at best, analogous effectiveness to traditional reading methods.

- **Clinical Decision Support**: AI systems show promise in supporting clinical decision-making processes. For example, a study presented an AI Clinician that learned optimal treatment strategies for sepsis (a life-threatening condition caused by the body's response to infection) in intensive care settings (Komorowski et al., 2018). The system analyzed data from patient admissions and demonstrated the potential to reduce patient mortality when its recommendations were followed. Similarly, in the field of diabetic retinopathy (a complication of diabetes that affects the eyes), a deep learning model[1] developed by Ryu et al. (2021) showed high accuracy in identifying the condition using optical coherence tomography angiography images (imaging technique to view blood vessels in the eye). The model showcased AI's potential to assist ophthalmologists in the early detection and management of diabetic eye complications.
- **Research and Biomedical Innovation**: Generative AI is also accelerating biomedical research and drug discovery processes (Cirillo et al., 2020a; Fleming, 2018; Poplin et al., 2018). For example, several authors discussed the case of Insilico Medicine, a generative AI-driven biotechnology company headquartered in Hong Kong. In less than a month, the company developed INS018_055, the first AI-designed drug for idiopathic pulmonary fibrosis, a progressive lung disease (Pun et al., 2023). This breakthrough demonstrates the efficiency of AI in drug development, significantly reducing the usual multi-year process from concept to initial trials to a much shorter timeframe.
- **Workflow Optimization**: Other authors argued that AI applications could alleviate managerial burdens and optimize clinical workflows (Bargalló et al., 2013; Riddleberger, 2024). This improvement includes assistance with tasks like medical coding, billing, and routine monitoring of symptoms (Velazquez & Florez, 2022). In clinical trials, studies suggest that AI is invaluable for patient recruitment and simulation. The technology can create digital representations of patients based on data from past trials and observational studies, potentially increasing confidence in results from smaller trials. This approach seems to be considered particularly beneficial for rare disease research, where the global patient population may be limited (Schott et al., 2015).

Challenges and concerning issues

This review also examined key issues surrounding generative AI in healthcare, focusing on privacy, algorithmic bias, and regulatory challenges. Recent studies explored how these factors intertwine, highlighting the need for enhanced security measures, awareness of inherent biases, and updated regulatory frameworks (Cirillo et al., 2020b, 2020a; Gerke et al., 2020; Sperling, 2019).

- **Data Privacy and Security**: Recent research emphasized the critical need to address security and privacy threats associated with generative AI in healthcare to ensure its safe and effective use (Chen & Esmaeilzadeh, 2024).

For example, some studies highlighted the multifaceted privacy challenges in healthcare AI (Marotta & Madnick, 2022). One emerging point of discussion is the issue of de-identification, which involves removing personal identifiers from data to protect privacy while preserving analytical value (Price & Cohen, 2019). Some authors pointed out that even with current de-identification techniques, there is a significant risk of re-identification when AI models are trained on large datasets (Kim et al., 2019; Rothstein, 2010). This risk is particularly pronounced with generative AI models that can potentially recreate detailed patient profiles. In terms of security, a comprehensive review by Seh et al. (2020) examined the vulnerabilities of healthcare systems to cyber-attacks. Along this line, other experts noted that integrating AI systems increases the attack surface, potentially exposing sensitive patient data to new forms of cyber threats (Guembe et al., 2022; Oseni et al., 2021). Some authors also attempted to find solutions to some of the challenges explained above. For instance, Rieke et al. (2020) proposed federated learning to address privacy concerns (i.e., training AI models on decentralized devices without sharing data), but noted this solution still involves security issues.

- **Algorithmic Bias and Fairness**: Researchers stressed the need to address biases in AI, which can lead to unfair and potentially risky outcomes in healthcare. For example, Obermeyer et al. (2019) found a widely used algorithm underestimated Black patients' care needs, worsening health disparities. Several authors also argued that gender biases pose significant risks in women's healthcare (Dhruva et al., 2011; Sperling, 2019). One key concern is that medical data used to train AI often reflect a male-dominated perspective, leading to misdiagnoses and delays in women's care. Marotta (2022) highlighted that biased AI can perpetuate the exclusion of women in healthcare, raising concerns about accountability and the need for better algorithmic audits.
- **Clinical Validity, Reliability, Explainability, and Transparency**: Ensuring clinical validity and reliability of AI insights requires rigorous testing and validation to build trust among healthcare professionals and patients. However, the "black box" nature of some AI algorithms poses challenges to explainability and transparency in medical decision-making (Nicholson Price, 2018; Yu & Ali, 2019). This lack of interpretability can complicate clinical adoption and raise ethical concerns. Goodman et al. (2018) explored the complexities of machine learning in healthcare, emphasizing the need for causal reasoning and transparency in AI models.
- **Regulatory and Compliance Issues**: Most authors argued that the rapid advancement of AI in healthcare is outpacing current regulatory frameworks (Hwang et al., 2019; Maresova, 2023; Van Laere et al., 2022). Studies agree that there is a need for updated policies and ethical guidelines to govern the development and deployment of these technologies. For example, McKee and Wouters (2023) highlighted the significant challenges in regulating

healthcare AI, noting that existing regulatory mechanisms are inadequate for dealing with the unique characteristics of AI applications.

Thus, generative AI holds significant potential to transform healthcare through innovations in personalized medicine, diagnostics, clinical decision support, and research. However, realizing these benefits requires careful navigation of challenges related to data privacy, algorithmic bias, clinical validity, and ethical considerations. This literature review established the foundation for evaluating the current healthcare landscape, with the goal of uncovering potential areas for advancement.

Methodology

This study employed a mixed-method approach to analyze the impact of AI technologies in healthcare. Anchored in a systematic literature review, the research design centered on a case study of HCB, complemented by qualitative insights from healthcare professionals and quantitative data from their empirical results.

- **Research Instrument**: A rigorously structured case study was employed to explore the application of generative AI at HCB. The hospital's pioneering role in implementing AI-based systems, particularly in radiology, made it an ideal site for examining the integration of advanced technologies. HCB's AI initiatives align with existing literature on AI's role in enhancing diagnostic precision and personalized care, further reinforcing the hospital's suitability for this case study.
- **Participants**: A purposive sampling method, which selects participants based on their expertise and relevance to the study's objectives, was adopted to identify key personnel with specialized knowledge of the AI system's implementation. This targeted approach was crucial for obtaining in-depth insights into its broader implications within the healthcare setting.
- **Procedure**: The case study examined the implementation process, benefits, and challenges. It integrated document analysis and qualitative insights from personnel involved in AI deployment. Data triangulation ensured reliability by cross-verifying sources. To complement the case study, information was obtained through a semi-structured interview guide, specifically crafted to elicit detailed responses on the AI system's impact. Open-ended questions were included to provide flexibility while ensuring key themes were thoroughly explored. Finally, as part of this comprehensive approach, a risk-benefit analysis using quantitative and qualitative data provided actionable insights for future AI implementations.

The research contributes to a nuanced understanding of AI's potential in healthcare delivery and decision-making by synthesizing case study evidence, expert perspectives, and contextual knowledge from the literature.

Case study: hospital clínic de Barcelona's journey into AI-driven healthcare applied to medical imaging

The following case study describes HBC's journey in integrating AI into clinical practice, marking a paradigm shift in patient care and diagnostic precision within the evolving medical landscape.

Background

HCB, with over a century of history, has long been a beacon of medical excellence, not just for Barcelona and Catalonia but for Spain and beyond. As a public healthcare provider, it serves a population of 540,000 within the Integrated Health Area "Barcelona Esquerra" (Área Integral de Salud de Barcelona Esquerra—AIS-BE) and acts as a national referral center for complex medical cases (*Hospital Clínic de Barcelona*, 2023). For decades, HCB has upheld a strong commitment to healthcare, research, and education. Now, the hospital is embracing a new era, integrating cutting-edge technology to meet the evolving demands of modern medicine.

Faced with the growing need for faster and more accurate diagnostics, HCB began exploring AI as a solution capable of not just meeting but exceeding these demands. According to Dr. Salvador Pedraza Gutierrez, Director of the Medical Imaging Department, this was a natural extension of the hospital's tradition of innovation: "*Half of our physicians have gone on sabbaticals to other hospitals, gaining experience in different scenarios and cultures*," Dr. Pedraza explained. This exposure to international practices fosters a rich, innovative environment that allows HCB to remain at the forefront of medical advancements. He described the hospital as "*a constellation of institutions*"—a collaborative network of specialties and research initiatives that work in harmony to drive progress.

In recent years, HCB has become a leader in AI implementation within healthcare. Reflecting on this journey, Dr. Pedraza remarked, "*Our AI initiatives began several years ago. Each step, whether successful or not, has been a valuable learning experience in our progression.*" This proactive approach to AI integration underscores the hospital's determination to remain at the cutting edge of medical technology. One of HCB's pioneering efforts in AI began with the adoption of Computer-Aided Detection (CAD)[2] systems in their breast cancer screening program. This initiative sought to enhance the accuracy and efficiency of mammography readings, addressing challenges, such as missed diagnoses due to low disease prevalence, complex breast density, subtle findings, and radiologist fatigue. For example, the CAD system showed promise in flagging subtle abnormalities in mammograms, potentially complementing the expertise of human radiologists. In the study conducted by a team of experienced radiologists, primarily from the Department of Radiology at HCB, CAD correctly marked 11 out of 14 cases that were initially read as negative by one of the radiologists (Bargalló et al., 2013). This research showcased AI's capability as an aiding resource in the diagnostic process and demonstrated HCB's commitment to taking charge of technological advancements.

As Dr. Pedraza pointed out, HCB's embrace of AI is not limited to clinical care. The hospital is also a leader in cybersecurity and data protection, with dedicated legal staff and a culture of continuous improvement, ensuring that patient care is both innovative and secure. Through its forward-thinking approach, HCB is not only treating patients but shaping the future of healthcare.

Implementation and evaluation of AI systems at HCB

Building upon the foundation of innovation and excellence, HCB's integration of AI technology has yielded both results and difficulties. This section critically examined these aspects of this implementation, drawing from detailed sources and critical interview evaluations.

Benefits of AI integration at HCB

At HCB, AI has become an integral part of a broader vision for the future of healthcare. For Dr. Salvador Pedraza, AI is not just a tool but a lifeline for Europe's healthcare systems, which are struggling to keep pace with the demands of an aging population, rising costs, and dwindling numbers of healthcare professionals. "*Our healthcare system is the most valuable part of our welfare in Europe*," he stated.

Therefore, as the pressure on healthcare systems became increasingly significant, HCB saw AI as the solution to alleviate some of the most burdensome challenges. Time is a critical resource in hospitals, and AI's ability to automate routine tasks quickly became a game-changer. "*We need to save time*," Dr. Pedraza emphasized. "*By automating easy processes, we can give back that time to our professionals, allowing them to dedicate themselves to more advanced and critical work, like the application of biomarkers*." As measurable indicators of biological states, biomarkers are essential for early disease detection, selecting treatments, and tracking patient responses. Therefore, with the help of AI to analyze complex biomarker data, HCB sought to improve diagnostic precision and customize treatments for each patient.

This notion of "giving back time" is one of the most tangible benefits of AI at HCB. One of the hospital's ventures into AI was through AI-powered chatbots designed to assist with appointment management. These chatbots quickly proved their worth by tackling a common issue in healthcare: patient no-shows. Dr. Pedraza described the persistent problem: "*You schedule an examination, but the patient doesn't come. This happens in any center—public or private—and results in a waiting list, wasting valuable machine time. It's a real problem*." To address this issue, the hospital launched a pilot study using a chatbot for densitometry (bone density measurement) appointments, where patients could easily provide information and reschedule appointments if needed. "*With the chatbot, we managed to reduce absenteeism by 50%. Now, we're expanding to mammography (breast X-ray imaging) and sonography (ultrasound imaging) appointments,*" he explained.

While the chatbot cannot entirely eliminate the problem, it has significantly reduced absenteeism. "*It's not about eliminating no-shows completely because people can still have unexpected issues—traffic accidents, medical problems—but we've reduced the rate,*" he noted. The key benefit, he highlighted, is how the technology empowers patients: "*The patient has the information and can say, 'I cannot come, please give me another appointment.' This is a simple application of technology, but it makes a big difference.*" By addressing something as straightforward as appointment scheduling, AI enables the hospital to better allocate resources, meaning more patients can be seen and treated on time.

Similarly, in terms of clinical outcomes, AI has demonstrated its potential to assist in improving cancer detection rates, as highlighted in studies performed by Dr. Xavier Bargalló, a physician who specialized in Diagnostic Imaging at HBC, and other colleagues (Bargalló et al., 2013, 2014). By integrating CAD with full-field digital mammography (FFDM)[3], the hospital saw improvements in cancer detection rates, with some studies reporting increases of up to 16%–20% (Bargalló et al., 2013, 2014). Dr. Bargalló's research showed that AI could support radiologists' work by automating routine tasks. This automation not only saved them valuable time but also enhanced overall detection sensitivity, particularly in identifying calcifications and small invasive breast cancers (Bargalló et al., 2013, 2014).

However, in addition to improving cancer detection, CAD emerged as a supportive tool in breast cancer screening. Dr. Bargalló and his team compared a novel screening approach—where a single expert radiologist used CAD—against the traditional method of double reading with arbitration (Bargalló et al., 2014). This innovative strategy demonstrated effectiveness comparable to (but not superior to) the established double-reading approach[4].

AI has also fostered a culture of continuous improvement at the hospital. Radiologists now undergo yearly performance reviews and participate in ongoing medical education, ensuring they remain on the cutting edge of both their profession and the AI technologies they use daily. "*It's not just about using AI—it's about evolving with it,*" says Dr. Pedraza. "*The more we learn, the better we become.*" For HCB, AI isn't just a technology—it's a partnership, one that is already transforming the way the hospital delivers care. "*AI has allowed us to be more efficient, more precise, and more patient-centered,*" Dr. Pedraza concluded.

Risks and challenges of AI at HCB

The implementation of AI at HCB presented a double-edged sword: while it promised revolutionary advancements in healthcare, it simultaneously introduced challenges that demanded meticulous attention. Dr. Pedraza's and his colleagues' insights revealed a complex landscape where the potential benefits of AI are constantly weighed against risks associated with security, compliance, patient safety, privacy, and trust.

At the forefront of these challenges is cybersecurity. Hospitals like HCB are prime targets for cyberattacks, jeopardizing sensitive medical data. This reality has prompted a thorough reevaluation of data security protocols and compliance with evolving regulations. Dr. Pedraza emphasizes, "*Safeguarding sensitive medical information is essential as we integrate AI, or any other technology, into our healthcare processes. We must prevent data breaches and ensure patient privacy remains intact.*" He underscored the necessity of precautions, noting, for example, the importance of limiting *"the use of external devices to maintain security and protect against potential risks.*"

Equally critical is the issue of validation in AI implementation. Dr. Pedraza stressed, "*In medicine, it's essential to conduct validation studies before using AI; evaluation must always come first.*" This consideration highlights a key issue in AI adoption: the requirement for comprehensive validation often conflicts with AI development's fast-paced nature. This concern is exemplified in the implementation of CAD systems for breast cancer screenings at HCB. While CAD improved cancer detection rates, it also led to an increase in the hospital's recall rate (RR, the percentage of patients called back for additional testing)—from 3.94% (using the traditional double reading plus arbitration method) to 7.02% (using the new single reading with CAD method) (Bargalló et al., 2014). Therefore, the increased RR resulted in a higher number of patients being summoned for follow-up examinations. During these additional assessments, radiologists reviewed the marked areas and rendered final decisions based on both the CAD-generated marks and their professional judgment.

Therefore, integrating AI in healthcare raises complex ethical and practical questions regarding responsibility and decision-making. To illustrate this point, Jorge Fernandez, head of innovation at HCB, highlighted a critical concern often overlooked in discussions about AI in healthcare: "*There's a real risk that radiologists, and doctors in general, might stop thinking critically and simply accept what the AI system suggests.*" As a result, this shift towards over-reliance on technology could lead to "accommodative attitudes" among health professionals. In this regard, Dr. Bargalló's statement, "*AI works very well, but it needs validation from a radiologist,*" highlights the importance of maintaining critical engagement when using AI in clinical practice (iSanidad, 2023). Building on this concept, Dr. Pedraza emphasized the ultimate accountability of medical professionals: "*The physician always bears the responsibility. It is always the physician who determines whether a finding is accurate or not.*" These expert opinions underscored the current state of AI in radiology—a powerful diagnostic tool that nonetheless requires human oversight due to inherent limitations. For instance, the CAD system exemplifies these challenges, demonstrating lower sensitivity for certain types of lesions in breast cancer screening, particularly architectural distortions, with a sensitivity of only 57.14% (Bargalló et al., 2013). This example means that CAD failed to detect nearly half of the architectural distortions present in mammograms, potentially missing important indicators of breast cancer. Moreover, the high rate of false-positive results generated by AI has, in some instances, increased radiologists' workload instead of alleviating it (Bargalló et al., 2014).

The issue of false positives extended to patient understanding and trust, adding another layer of complexity. For example, AI-generated markings on medical images can potentially cause a critical communication gap between healthcare professionals and patients. Dr. Pedraza observed, "*AI is not a perfect tool—it can make mistakes, such as false positives.*" These errors, though typically addressed by physicians, could worry patients. "*Patients may not fully grasp the limitations of AI, and that can lead to unnecessary confusion,*" he warned.

Navigating the regulatory landscape presented another significant challenge in AI implementation. The rapid evolution of AI technology is outpacing regulatory frameworks, creating an uncertain environment where balancing innovation and compliance becomes a delicate act. Despite these hurdles, HCB's openness to change has been fundamental in navigating these challenges. Dr. Pedraza highlighted the legal team's adaptability, stating, "*We're fortunate that our legal team is very open to adapting and implementing these new legal rules in our hospital.*" In particular, the Director of Data Protection played a key role in this context, overseeing all compliance issues and serving as the go-to person for consultations.

Emphasizing the critical nature of regulatory compliance, Jorge Fernandez highlighted, "*If an AI algorithm doesn't meet regulations, regardless of whether we developed it internally or acquired it from a company, we can't use it for patient care.*" This comment further underscored the importance of ensuring all AI technologies comply with relevant regulations before they can be implemented in clinical practice. However, Dr. Pedraza also acknowledged the difficulties in aligning medical professionals' expertise with regulatory requirements. He explained:

> It can be challenging because medical staff highly value their professional independence. But rather than seeing legal safeguards as restrictive, we've found that fostering open communication is necessary. We're working to help our staff recognize that these regulations are essential for maintaining high standards and preventing potential issues.

Therefore, integrating AI into HCB's practices required balancing innovation with risk mitigation. The interviewees' insights revealed that successful implementation hinges not just on technical advancement, but also on managing human factors, ethical considerations, and overall institutional preparedness.

This readiness also extends to financial capacity, a critical component of an organization's ability to adopt and sustain AI. As Fernandez emphasized, "*The ability to afford AI technologies is critical. Without proper funding, hospitals risk falling behind.*" Considering these constraints, HCB's AI-driven success depends on effectively addressing these multifaceted challenges, ensuring AI benefits are realized while upholding patient care principles and maintaining the hospital's innovative edge.

Results: a risk-benefit analysis of the dual nature of AI in healthcare at hospital clínic de Barcelona

This research examined the AI implementation at HCB through a risk-benefit analysis of key operational areas highlighted in the case study. A 3×3 risk matrix was then applied to capture the nuanced balance of benefits and challenges (Table 12.1).

This matrix evaluates each AI factor impacting a given domain based on its probability (likelihood of occurrence) and impact (potential consequences), both rated on a scale from 1 (low) to 3 (high). The risk score, calculated by multiplying probability and impact values, ranges from 1 to 9 and is categorized into four risk levels: low (1–2), moderate (3–4), high (6), or critical (9). In the analysis, positive factors (benefits) and negative factors (risks) were evaluated separately using this matrix. The prevailing outcome in each domain was determined by comparing benefit scores against risk scores. When benefit scores exceeded risk scores, benefits were considered to prevail, and vice versa. In the case of equal scores, additional factors, such as the broader context and potential long-term implications, were factored in to determine the overall impact. This approach allowed for a nuanced understanding of AI's potential outcomes in healthcare, considering both quantitative data and qualitative assessments to evaluate its implementation comprehensively.

The analysis of each domain can be understood within this framework:

- **Diagnostic Accuracy and Efficiency**: HCB's integration of CAD systems with FFDM exemplifies a structured approach to assessing the risk-benefit

Table 12.1 Risk matrix

Probability	*Impact severity*		
	Low (1)	*Medium (2)*	*High (3)*
High (3)	3 (Moderate risk)	6 (High risk)	9 (Critical risk)
Medium (2)	2 (Low risk)	4 (Moderate risk)	6 (High risk)
Low (1)	1 (Low risk)	2 (Low risk)	3 (Moderate risk)

Source: Own elaboration

ratio. Studies involving large numbers of screening mammograms provided quantitative data on performance improvements (Bargalló et al., 2013, 2014). According to the matrix, the increased cancer detection rate scores 9 (High Impact: 3, High Probability: 3), reflecting its critical positive impact. The increased recall rate scores 6 (Medium Impact: 2, High Probability: 3), indicating a high but manageable risk. This quantitative approach supports the qualitative assessment that benefits outweigh risks. Additionally, the evaluation considers long-term health outcomes, the potential for early intervention, and the ability to detect difficult-to-identify lesions, further tilting the balance towards the advantages of AI integration in this healthcare domain.

- **Time Management and Workflow**: HCB's implementation of AI-powered chatbots for appointment management demonstrated significant benefits. The 50% reduction in absenteeism for certain procedures indicates substantial operational efficiency gains. The risk assessment assigned a score of 6 to reduced absenteeism (High Impact: 3, Medium Probability: 2) compared to a score of 4 for increased workload (Medium Impact: 2, Medium Probability: 2). This quantitative analysis supports the qualitative assessment that the benefits of improved resource utilization and care delivery efficiency outweigh the manageable risks associated with potential workload fluctuations due to false-positive results in diagnostic imaging.
- **Patient Engagement and Communication**: HCB's AI implementation in this domain presents a complex landscape. While AI significantly enhances information accessibility and appointment efficiency (Medium Impact: 2, Medium-High Probability: 2.5, Score: 5), the slightly higher risk of misinterpreting AI-generated data, particularly in critical areas like medical imaging (High Impact: 3, Medium Probability: 2, Score: 6), cannot be overlooked. This nuanced risk-benefit ratio argues for a cautious approach, emphasizing that the potential impact on patient understanding and trust is paramount despite quantifiable improvements.
- **Data Security and Regulatory Compliance**: HCB's proactive approach to data security and regulatory compliance, exemplified by appointing a Director of Data Protection, demonstrates a solid commitment to risk mitigation. However, the rapidly evolving nature of AI technology and legal frameworks presents significant challenges. While enhanced data protection measures score highly (6: High Impact 3, Medium Probability 2), the equally high score for cybersecurity threats (6: High Impact 3, Medium Probability 2) underscores the persistent risks. However, despite HCB's comprehensive strategy, this domain's inherent complexity and dynamic nature suggest that risks may outpace benefits without continuous, vigilant adaptation of security protocols and compliance measures. The analysis compels a cautious perspective, emphasizing that even robust current measures may struggle to keep pace with emerging threats and regulatory changes in the AI-driven healthcare landscape.

- **Clinical Decision-Making and Professional Autonomy**: The integration of AI at HCB presented a nuanced scenario. While AI enhances analytical capabilities and potentially improves diagnostic accuracy, it also risks eroding professional autonomy. The risk assessment scored improved diagnostic accuracy at 6 (High Impact: 3, Medium Probability: 2) and erosion of professional autonomy at 2 (Medium Impact: 2, Low Probability: 1). This significant difference in scores indicates that the benefits of AI in clinical decision-making outweigh the risks. However, HCB's emphasis on validation studies and maintaining physician responsibility demonstrates an awareness of the need to balance AI's analytical power with preserving medical professionals' critical thinking and decision-making skills.

Discussion: recommendations for navigating the AI paradox in healthcare

The study at HCB addressed the original research question: Under what conditions can AI enhance or hinder healthcare progress? More specifically, the research revealed that AI can significantly improve healthcare when:

- It complements human expertise rather than replacing it, as seen in improved cancer detection rates with CAD systems.
- Implementation prioritizes transparency, minimizing bias in AI systems.
- Robust cybersecurity measures and ethical guidelines are in place to protect patient data and trust.
- Comprehensive staff training and organizational readiness precede AI integration.

Conversely, AI can hinder progress when:

- It compromises professional autonomy or patient trust.
- Implementation leads to over-reliance on technology, potentially increasing false positives or recall rates.
- Patient education about AI-generated data is insufficient, risking misinterpretation.
- Data security and privacy concerns are not adequately addressed.

The study concluded that successful AI integration in healthcare requires a balanced approach that maximizes benefits while mitigating risks. This outcome involves continuous assessment, adaptive strategies, and a focus on maintaining the human element in patient care. The HCB case demonstrated that when these conditions are met, AI can significantly enhance diagnostic capabilities, workflow efficiency, and overall healthcare delivery. At the same time, careful management is crucial to navigate challenges and ensure responsible AI adoption.

Table 12.2 Benefits, challenges, and areas for improvement

Domains	*Benefits*	*Challenges*	*Prevailing aspect*	*Areas for improvement*
Diagnostic Accuracy and Efficiency	Improved cancer detection rates	Increased recall rates	Benefit	AI Algorithm Refinement
Time Management and Workflow	Automation of routine tasks, reduction in no-shows	Potential increase in radiologists' workload due to false positives	Benefit	Adaptive Workflow Integration
Patient Engagement and Communication	Enhanced appointment management, improved access to information	Potential misunderstandings of AI-generated data	Challenge	Comprehensive Patient Education
Data Security and Regulatory Compliance	Proactive cybersecurity measures	Rapidly evolving technology outpacing regulatory frameworks	Challenge	Dynamic Security and Regulatory Adaptation
Clinical Decision-Making and Professional Autonomy	Advanced analytical capabilities, particularly in personalized medicine	Balancing AI assistance with professional independence	Benefit	AI-Clinician Collaborative Model

Source: Own elaboration

To better explain these considerations, the following table (Table 12.2) provides a clearer picture of AI's impact at HCB, highlighting areas where benefits prevail and where challenges require focused attention. The column "Areas for Improvement" suggests targeted strategies to address prevailing challenges and further enhance benefits, ensuring healthcare facilities like HCB can maximize the potential of AI in healthcare while effectively managing associated risks.

The identified areas for improvement in AI implementation at HCB address critical aspects of healthcare delivery and technology integration. These improvements are essential for maximizing the benefits of AI while mitigating associated challenges:

- **AI Algorithm Refinement**: This enhancement aims to improve the accuracy of AI systems in diagnostics, particularly in breast cancer screening. By adopting advanced algorithms, HCB can effectively reduce false positives while maintaining high sensitivity for disease detection. To achieve this, techniques such as deep learning—where neural networks learn patterns from large datasets—and ensemble methods, which combine the predictions of multiple models to improve accuracy, could be employed to optimize AI performance based on ongoing data analysis. To support this optimization, key strategies may include conducting regular algorithm audits and collaborating with various medical institutions to enrich training datasets. Furthermore, creating feedback loops for radiologists to report AI errors will enhance the overall system. Together, these improvements can specifically address the challenge of high recall rates in breast cancer screening, ultimately leading to more efficient resource utilization and reducing patient anxiety caused by unnecessary follow-ups.

- **Adaptive Workflow Integration**: Establishing flexible systems that optimize AI-assisted processes without overburdening staff can be essential for unlocking AI's potential in healthcare. This improvement aims to integrate AI tools into existing workflows, ensuring that automation saves time while enhancing patient care rather than increasing administrative tasks. Key steps in this process can include establishing standardized data formats across departments for compatibility, implementing secure Application Programming Interfaces for efficient data exchange, and developing a centralized data lake—a repository that consolidates comprehensive patient information. Additionally, intelligent task allocation systems can prioritize and distribute workloads based on urgency, balancing the advantages of automation with the necessity of human expertise. This approach effectively addresses radiologists' increased workloads due to AI-generated false positives, ensuring that AI streamlines healthcare delivery. Moreover, the HCB case study emphasizes the importance of change management in AI implementation. Successful integration requires not only technological updates but also cultural shifts within the organization. This includes developing AI literacy programs to educate staff, conducting workshops to inform clinical teams about advancements, and establishing mentorship initiatives that connect AI experts with clinical personnel.
- **Comprehensive Patient Education**: Implementing robust programs to enhance understanding of AI in healthcare is essential for reducing patient misinterpretation of AI-generated data. This initiative should focus on providing clear, accessible explanations of AI's role, capabilities, and limitations in healthcare. To achieve this improvement, key strategies may involve creating informative videos visually depicting how AI functions in medical contexts, developing interactive apps that enable patients to explore and understand AI-assisted diagnoses, and establishing feedback systems that allow patients to voice concerns about AI applications.
- **Dynamic Security and Regulatory Adaptation**: From a cybersecurity perspective, integrating AI systems into HCB's existing infrastructure introduces new and evolving challenges. As healthcare institutions increasingly rely on AI and digital technologies, robust cybersecurity measures are essential to

protect sensitive patient data and ensure the integrity of AI-assisted diagnostic processes. Additionally, since security and compliance are closely intertwined, establishing agile regulatory approaches is crucial for keeping pace with AI advancements while ensuring adherence to legal standards. This improvement could involve creating a dedicated AI security and compliance team that collaborates with regulatory bodies to develop flexible guidelines. Such an approach would enable healthcare facilities to innovate more freely while maintaining strong data protection and ethical standards.

- **AI-Clinician Collaborative Model**: Developing AI tools that support clinical decision-making while preserving professional judgment and autonomy is essential for practical integration into healthcare. A viable approach that healthcare facilities can implement is the "Human-in-the-Loop" AI system (Zhou et al., 2022). This system allows clinicians to actively refine and validate AI outputs, ensuring that human expertise informs AI recommendations. With a real-time feedback mechanism, clinicians can flag, annotate, or correct AI suggestions immediately as they arise. By involving healthcare professionals in the AI development process, organizations can create tools that enhance human decision-making rather than replace it. Additionally, ongoing training for medical staff on how to effectively utilize AI insights can further strengthen this collaboration, promoting a harmonious relationship between AI technology and clinical expertise.

These areas of improvement are linked and mutually reinforcing. For instance, AI algorithm refinement directly supports adaptive workflow integration by providing more accurate and reliable AI outputs. Similarly, comprehensive patient education complements the AI-clinician collaborative model by ensuring that patients understand and trust the role of AI in their care. By focusing on these improvements, HCB and other healthcare facilities can create a more robust and effective AI ecosystem within their healthcare environment. This approach not only addresses current challenges but also positions hospitals to better leverage future advancements in AI technology.

Conclusion

This research makes significant theoretical and practical contributions to AI in healthcare. Theoretically, the results from HCB's AI implementation align with the broader narrative of AI in healthcare, showcasing both its potential and significant risks. For example, the enhanced diagnostic capabilities in radiology demonstrate how AI can improve patient outcomes through faster and more accurate diagnoses. This consideration aligns with previous research predictions about AI's ability to advance medical imaging. Similarly, the challenges encountered, particularly in data security and the potential for reduced human interaction in patient care, highlight the need to review existing mechanisms for dealing with the unique features of AI systems. However, these findings also contribute new insights to the ongoing dialogue about AI in healthcare, underlining that AI's success in healthcare depends on conditional factors to maximize its benefits. Additionally, it also stresses the ongoing need for vigilant oversight and adaptive strategies to address emerging challenges, thus contributing to the broader discourse on responsible AI adoption in healthcare.

From a practical standpoint, this research provides valuable insights for those contemplating AI integration in healthcare. It offers a comprehensive framework for assessing the benefits and risks associated with AI implementation, emphasizing the necessity of a holistic approach. Therefore, the study's recommendations can serve as a strategic guide for healthcare leaders, compliance experts, and technology developers, emphasizing both the positive potential of AI and the critical considerations essential for its effective integration into healthcare systems.

However, the case of HCB also raises important questions about the scalability and generalizability of AI solutions in healthcare, which may also represent a possible limitation of this study. While the implementation at HCB shows promising results, it is crucial to consider how such solutions might be adapted and implemented in diverse healthcare settings with varying resources, patient populations, and regulatory environments. This consideration is essential for understanding the broader implications of AI in healthcare and its capability to address health disparities. Therefore, future research could benefit from expanding the scope to include diverse healthcare settings and cultural contexts, potentially offering even richer insights into the global applicability of AI in medicine. Another critical area for future research lies at the intersection of AI implementation, compliance, and cybersecurity in healthcare. As AI systems become more deeply integrated into healthcare operations, novel data protection and regulatory compliance challenges are likely to emerge. Further studies could explore innovative approaches to ensuring AI systems' compliance with evolving healthcare regulations while maintaining robust cybersecurity measures. These studies might involve investigating AI-powered compliance monitoring systems or developing new frameworks for assessing and mitigating AI-specific cybersecurity risks in healthcare environments.

In conclusion, the implications of this study extend beyond the immediate context of AI in diagnostics. They touch upon broader themes of healthcare transformation and underscore the need for interdisciplinary collaboration in shaping the future of AI in healthcare, involving not just medical professionals and AI experts but also cybersecurity professionals and policymakers.

Notes

1 Advanced AI techniques used to refine diagnostic algorithms.
2 CAD is a multidisciplinary technology that combines AI and computer vision with the processing of radiological and pathology images. It is designed to create methods that help physicians interpret medical images, provide a second opinion, and assist in making final diagnostic decisions (Sampat et al., 2005; Yanase & Triantaphyllou, 2019).

3 A modern breast imaging technology used at HCB that creates digital images of the breast, allowing for better image manipulation and storage compared to traditional film-based mammography.
4 It's important to note that the success of this approach relied on selecting radiologists who are breast specialists or have a proven track record of good performance in mammography interpretation. This strategy is not suitable for radiologists from other specialties, such as abdominal radiologists, and should not be implemented merely for convenience.

References

AIPRM. (2024). 50+ AI in Healthcare Statistics 2024. https://www.aiprm.com/ai-in-healthcare-statistics/

Anderson, J. M., Tejani, T., Jarmain, B., Kellett, L., & Moy, R. L. (2023). Artificial intelligence vs medical providers in the dermoscopic diagnosis of melanoma. *Cutis, 111*(5), 254–258. https://doi.org/10.12788/cutis

Bargalló, X., Santamaría, G., Del Amo, M., Arguis, P., Ríos, J., Grau, J., Burrel, M., Cores, E., & Velasco, M. (2014). Single reading with computer-aided detection performed by selected radiologists in a breast cancer screening program. *European Journal of Radiology*, *83*(11), 2019–2023. https://doi.org/10.1016/j.ejrad.2014.08.010

Bargalló, X., Velasco, M., Santamaría, G., Del Amo, M., Arguis, P., & Sánchez Gómez, S. (2013). Role of computer-aided detection in very small screening detected invasive breast cancers. *Journal of Digital Imaging*, *26*(3), 572–577. https://doi.org/10.1007/s10278-012-9550-y

Bates, D. W., Auerbach, A., Schulam, P., Wright, A., & Saria, S. (2020). Reporting and implementing interventions involving machine learning and artificial intelligence. *Annals of Internal Medicine*, *172*(11), S137–S144. https://doi.org/10.7326/M19-0872

Bleher, H., & Braun, M. (2022). Diffused responsibility: attributions of responsibility in the use of AI-driven clinical decision support systems. *AI and Ethics*, *1*, 1. https://doi.org/10.1007/s43681-022-00135-x

Chen, Y., & Esmaeilzadeh, P. (2024). Generative AI in medical practice: in-depth exploration of privacy and security challenges. *Journal of Medical Internet Research*, *26*(1), e53008. https://doi.org/10.2196/53008

Cirillo, D., Catuara-Solarz, S., Morey, C., Guney, E., Subirats, L., Mellino, S., Gigante, A., Valencia, A., Rementeria, M. J., Chadha, A. S., & Mavridis, N. (2020a). Sex and gender differences and biases in artificial intelligence for biomedicine and healthcare. *npj Digital Medicine*, *3*(1), 1–11. https://doi.org/10.1038/s41746-020-0288-5

Cirillo, D., Catuara-Solarz, S., Morey, C., Guney, E., Subirats, L., Mellino, S., Gigante, A., Valencia, A., Rementeria, M. J., Chadha, A. S., & Mavridis, N. (2020b). Sex and gender differences and biases in artificial intelligence for biomedicine and healthcare. *npj Digital Medicine*, 3(1), 1–11. https://doi.org/10.1038/s41746-020-0288-5

Davenport, T., & Kalakota, R. (2019). The potential for artificial intelligence in healthcare. *Future Healthcare Journal*, *6*(2), 94–98. https://doi.org/10.7861/futurehosp.6-2-94

De Fauw, J., Ledsam, J. R., Romera-Paredes, B., Nikolov, S., Tomasev, N., Blackwell, S., Askham, H., Glorot, X., O'Donoghue, B., Visentin, D., van den Driessche, G., Lakshminarayanan, B., Meyer, C., Mackinder, F., Bouton, S., Ayoub, K., Chopra, R., King, D., Karthikesalingam, A., … Ronneberger, O. (2018). Clinically applicable deep learning for diagnosis and referral in retinal disease. *Nature Medicine*, *24*(9), 1342–1350. https://doi.org/10.1038/s41591-018-0107-6

Dhruva, S. S., Bero, L. A., & Redberg, R. F. (2011). Gender bias in studies for food and drug administration premarket approval of cardiovascular devices. *Circulation: Cardiovascular Quality and Outcomes*, 4(2), 165–171. https://doi.org/10.1161/CIRCOUTCOMES.110.958215

European Union. (2024). *Regulation (EU) 2024/1689 of the European Parliament and of the Council of 10 July 2024 on AI Compliance*. Luxembourg: Publications Office of the European Union.

Fleming, N. (2018). How artificial intelligence is changing drug discovery spotlight. *Nature*, *557*(7707), S55–S57. https://doi.org/10.1038/d41586-018-05267-x

Garbin, C., & Marques, O. (2022). Assessing methods and tools to improve reporting, increase transparency, and reduce failures in machine learning applications in health care. *Radiology: Artificial Intelligence*, *4*(2), 1–9. https://doi.org/10.1148/RYAI.210127

Gerke, S., Minssen, T., & Cohen, G. (2020). Ethical and legal challenges of artificial intelligence-driven healthcare. In A. Bohr & K. Memarzadeh (Eds.), *Artificial Intelligence in Healthcare* (pp. 295–336). Elsevier. https://doi.org/10.1016/b978-0-12-818438-7.00012-5

Ghebrehiwet, I., Zaki, N., Damseh, R., & Mohamad, M. S. (2024). Revolutionizing personalized medicine with generative AI: a systematic review. *Artificial Intelligence Review*, *57*(5), 1–41. https://doi.org/10.1007/s10462-024-10768-5

Goodman, S. N., Goel, S., & Cullen, M. R. (2018). Machine learning, health disparities, and causal reasoning. *Annals of Internal Medicine*, *169*(12), 883–885. https://www.acpjournals.org/doi/10.7326/M18-3297

Guembe, B., Azeta, A., Misra, S., Osamor, V. C., Fernandez-Sanz, L., & Pospelova, V. (2022). Applied artificial intelligence the emerging threat of AI-driven cyber attacks: a review. *Applied Artificial Intelligence*, *36*(1), e2037254. https://doi.org/10.1080/08839514.2022.2037254

Hospital Clínic de Barcelona. (2023). Memoria 2023. https://www.clinicbarcelona.org/asistencia/memoria-2023

Hu, Y., Jacob, J., Parker, G. J. M., Hawkes, D. J., Hurst, J. R., & Stoyanov, D. (2020). The challenges of deploying artificial intelligence models in a rapidly evolving pandemic. *Nature Machine Intelligence*, *2*(6), 298–300. https://doi.org/10.1038/s42256-020-0185-2

Hwang, T. J., Kesselheim, A. S., & Vokinger, K. N. (2019). Lifecycle regulation of artificial intelligence- and machine learning-based software devices in medicine. *JAMA – Journal of the American Medical Association*, *322*(23), 2285–2286. https://doi.org/10.1001/JAMA.2019.16842

iSanidad. (2023). Dr. Xavier Bargalló: La IA funciona muy bien, pero necesita de la validación de un radiólogo – YouTube [Broadcast]. *YouTube*. https://www.youtube.com/watch?v=wffy8Vngb8A

Kim, H. H., Kim, B., Joo, S., Shin, S. Y., Cha, H. S., & Park, Y. R. (2019). Why do data users say health care data are difficult to use? a cross-sectional survey study. *Journal of Medical Internet Research*, *21*(8), e14126. https://doi.org/10.2196/14126

Komorowski, M., Celi, L. A., Badawi, O., Gordon, A. C., & Faisal, A. A. (2018). The artificial intelligence clinician learns optimal treatment strategies for sepsis in intensive care. *Nature Medicine*, *24*(11), 1716–1720. https://doi.org/10.1038/s41591-018-0213-5

Lysaght, T., Lim, H. Y., Xafis, V., & Ngiam, K. Y. (2019). AI-assisted decision-making in healthcare: the application of an ethics framework for big data in health and research. *Asian Bioethics Review*, *11*(3), 299–314. https://doi.org/10.1007/s41649-019-00096-0

Maresova, P. (2023). Impact of regulatory changes on innovations in the medical device industry comment on "Clinical decision support and new regulatory frameworks for medical devices: are we ready for it?–a viewpoint paper." *International Journal of Health Policy and Management*, *12*(1), 1–4. https://doi.org/10.34172/IJHPM.2022.7262

Marotta, A. (2022). When AI is wrong: addressing liability challenges in women's healthcare. *Journal of Computer Information Systems*, *62*(6), 1310–1319. https://doi.org/10.1080/08874417.2022.2089773

Marotta, A., Madnick, B., & Madnick, S. (2023). Navigating the Wave of Cybersecurity Regulations: A Systematic Analysis of Emerging Regulatory Developments. *AMCIS 2023 Proceedings*, Panama City, Panama. https://aisel.aisnet.org/amcis2023/sig_sec/sig_sec/12

Marotta, A., & Madnick, S. (2022). Cybersecurity as a unifying factor for privacy, compliance and trust: the Haga hospital case. *Issues In Information Systems*, *23*(1), 102–116. https://doi.org/10.48009/1_iis_2022_108

McKee, M., & Wouters, O. J. (2023). The challenges of regulating artificial intelligence in healthcare: comment on "Clinical decision support and new regulatory frameworks for medical devices: are we ready for it? – a viewpoint paper." *International Journal of Health Policy and Management*, *12*(1), 7261. https://doi.org/10.34172/IJHPM.2022.7261

McKinney, S. M., Sieniek, M., Godbole, V., Godwin, J., Antropova, N., Ashrafian, H., Back, T., Chesus, M., Corrado, G. C., Darzi, A., Etemadi, M., Garcia-Vicente, F., Gilbert, F. J., Halling-Brown, M., Hassabis, D., Jansen, S., Karthikesalingam, A., Kelly, C. J., King, D., … Shetty, S. (2020). International evaluation of an AI system for breast cancer screening. *Nature*, *577*(7788), 89–94. https://doi.org/10.1038/s41586-019-1799-6

Mun, S. K., Wong, K. H., Lo, S. C. B., Li, Y., & Bayarsaikhan, S. (2021). Artificial intelligence for the future radiology diagnostic service. *Frontiers in Molecular Biosciences*, *7*, 512. https://doi.org/10.3389/fmolb.2020.614258

Nicholson Price, W. (2018). Big data and black-box medical algorithms. In *Science Translational Medicine* (Vol. 10, Issue 471). American Association for the Advancement of Science. https://doi.org/10.1126/scitranslmed.aao5333

Obermeyer, Z., Powers, B., Vogeli, C., & Mullainathan, S. (2019). Dissecting racial bias in an algorithm used to manage the health of populations. *Science*, *366*(6464), 447–453. https://doi.org/10.1126/science.aax2342

Oseni, A., Moustafa, N., Janicke, H., Liu, P., Tari, Z., & Vasilakos, A. (2021). Security and privacy for artificial intelligence: opportunities and challenges; security and privacy for artificial intelligence: opportunities and challenges. *arXiv* preprint arXiv:2102.04661. https://arxiv.org/abs/2102.04661

Poplin, R., Varadarajan, A. V., Blumer, K., Liu, Y., McConnell, M. V., Corrado, G. S., Peng, L., & Webster, D. R. (2018). Prediction of cardiovascular risk factors from retinal fundus photographs via deep learning. *Nature Biomedical Engineering*, *2*(3), 158–164. https://doi.org/10.1038/s41551-018-0195-0

Price, W. N., & Cohen, I. G. (2019). Privacy in the age of medical big data. *Nature Medicine*, *25*(1), 37–43. https://doi.org/10.1038/s41591-018-0272-7

Pun, F. W., Ozerov, I. V, & Zhavoronkov, A. (2023). AI-powered therapeutic target discovery. *Trends in Pharmacological Sciences*, *44*(9). https://doi.org/10.1016/j.tips.2023.06.010

Reddy, S. (2024). Generative AI in healthcare: an implementation science informed translational path on application, integration and governance. *Implementation Science*, *19*(27). https://doi.org/10.1186/s13012-024-01357-9

Riddleberger, K. (2024). Revolutionizing Healthcare: The Transformative Power of AI. American Academy of Physician Associates. https://www.aapa.org/news-central/2024/05/revolutionizing-healthcare-the-transformative-power-of-ai/

Rieke, N., Hancox, J., Li, W., Milletarì, F., Roth, H. R., Albarqouni, S., Bakas, S., Galtier, M. N., Landman, B. A., Maier-Hein, K., Ourselin, S., Sheller, M., Summers, R. M., Trask, A., Xu, D., Baust, M., & Cardoso, M. J. (2020). The future of digital health with federated learning. *Npj Digital Medicine*, *3*(119). https://doi.org/10.1038/s41746-020-00323-1

Rong, G., Mendez, A., Bou Assi, E., Zhao, B., & Sawan, M. (2020). Artificial intelligence in healthcare: review and prediction case studies. *Engineering*, *6*(3), 291–301. https://doi.org/10.1016/j.eng.2019.08.015

Rothstein, M. A. (2010). Is deidentification sufficient to protect health privacy in research? NIH public access. *The American Journal of Bioethics*, *10*(9), 3–11. https://doi.org/10.1080/15265161.2010.494215

Ryu, G., Lee, K., Park, D., Park, S. H., & Sagong, M. (2021). A deep learning model for identifying diabetic retinopathy using optical coherence tomography angiography. *Scientific Reports*, *11*(1), 1–9. https://doi.org/10.1038/s41598-021-02479-6

Saeed, M., Naseer, A., Masood, H., Rehman, S. U., & Gruhn, V. (2023). Smart healthcare in the era of artificial intelligence: a systematic review. *IEEE Access*, *11*, 34677–34694. https://arxiv.org.

Salinas, M. P., Sepúlveda, J., Hidalgo, L., Peirano, D., Morel, M., Uribe, P., Rotemberg, V., Briones, J., Mery, D., & Navarrete-Dechent, C. (2024). A systematic review and meta-analysis of artificial intelligence versus clinicians for skin cancer diagnosis. *Npj Digital Medicine*, *7*(1), 1–23. https://doi.org/10.1038/s41746-024-01103-x

Sampat, M. P., Markey, M. K., & Bovik, A. C. (2005). Computer-aided detection and diagnosis in mammography. In A. C. Bovik (Ed.), *Handbook of Image and Video Processing* (Second Edition, pp. 1195–1217). Burlington, MA: Elsevier Academic Press. https://doi.org/10.1016/B978-012119792-6/50130-3

Sanford, T., Harmon, S. A., Turkbey, E. B., Kesani, D., Tuncer, S., Madariaga, M., Yang, C., Sackett, J., Mehralivand, S., Yan, P., Xu, S., Wood, B. J., Merino, M. J., Pinto, P. A., Choyke, P. L., & Turkbey, B. (2020). Deep-learning-based artificial intelligence for PI-RADS classification to assist multiparametric prostate MRI interpretation: a development study. *Journal of Magnetic Resonance Imaging*, *52*(5), 1499–1507. https://doi.org/10.1002/jmri.27204

Schott, A. F., Welch, J. J., Verschraegen, C. F., & Kurzrock, R. (2015). The national clinical trials network: conducting successful clinical trials of new therapies for rare cancers. In *Seminars in Oncology* (Vol. 42, Issue 5, pp. 731–739). W.B. Saunders. https://doi.org/10.1053/j.seminoncol.2015.07.010

Schwendicke, F., & Krois, J. (2021). Better reporting of studies on artificial intelligence: CONSORT-AI and beyond. *Journal of Dental Research*, *100*(7), 677–680. https://doi.org/10.1177/0022034521998337

Seh, A. H., Zarour, M., Alenezi, M., Sarkar, A. K., Agrawal, A., Kumar, R., & Khan, R. A. (2020). Healthcare data breaches: insights and implications. *Healthcare*, *8*(2), 1–18. https://doi.org/10.3390/HEALTHCARE8020133

Shah, N. R. (2019). Health care in 2030: will artificial intelligence replace physicians? *Annals of Internal Medicine*, *170*(6), 407–408. https://doi.org/10.7326/M19-0344

Shaheen, M. Y. (2021). Applications of Artificial Intelligence (AI) in Healthcare: A Review. *ScienceOpen Preprints*. https://doi.org/10.14293/S2199-1006.1.SOR-.PPVRY8K.v1

Sperling, E. (2019). *How Hardware Can Bias AI Data*. Semiconductor Engineering. https://semiengineering.com/where-data-gets-biased/

U.S. Government Accountability Office. (2024). Generative AI in Health Care (GAO-24-107634). https://www.gao.gov/assets/880/871171.pdf

Van Laere, S., Muylle, K. M., & Cornu, P. (2022). Clinical decision support and new regulatory frameworks for medical devices: are we ready for it?-a viewpoint paper. *International Journal of Health Policy and Management*, *11*(12), 3159–3163. https://doi.org/10.34172/IJHPM.2021.144

Velazquez, A. I., & Florez, N. (2022). Electronic monitoring of patient-reported outcomes—ready for prime time? *JAMA Oncology*, *8*(8), 1099–1100. https://doi.org/10.1001/JAMAONCOL.2022.2927

Yanase, J., & Triantaphyllou, E. (2019). A systematic survey of computer-aided diagnosis in medicine: past and present developments. *Expert Systems with Applications*, *138*. https://doi.org/10.1016/j.eswa.2019.112821

Yu, R., & Alì, G. S. (2019). What's inside the black box? AI challenges for lawyers and researchers. In *Legal Information Management* (Vol. 19, Issue 01, pp. 2–13). Cambridge University Press (CUP). https://doi.org/10.1017/s1472669619000021

Zhou, T., Shen, J., He, D., Vijayakumar, P., & Kumar, N. (2022). Human-in-the-loop-aided privacy-preserving scheme for smart healthcare. *IEEE Transactions on Emerging Topics in Computational Intelligence*, *6*(1), 6–15. https://doi.org/10.1109/TETCI.2020.2993841

13 The use of generative AI in the enterprise—from uncertainty to trust

Wojciech Pizło

Introduction

Digitalization affects both private life and professional spheres. It takes place not only on a macro scale but, above all, on a micro scale, impacting individual enterprises (Pizło, 2009, pp. 75–108). Since the public release of "AI" software as a machine learning tool, there has been widespread interest in this technology and its capabilities. Entrepreneurs have focused more attention on Artificial intelligence (AI) as a relatively new tool capable of rationalizing many tasks, accelerating organizational processes, and increasing human productivity. Technological changes (the widespread availability of network access) contribute to the acquisition of Big Data, which allows for the identification of key processes occurring within organizations, as well as the identification of behaviors of individual market entities. Business activities undertaken with the use of AI are driven by increasing competitiveness. The aim of this work is to identify changes occurring in enterprises under the influence of AI.

Review of literature

Artificial Intelligence —definitions and its applications

AI is widely regarded as one of the most promising digital technologies. One of the most commonly accepted definitions of AI is by Kaplan and Haenlein (2019, p. 17), who described AI as "a system's ability to interpret external data correctly, to learn from such data, and to use those learnings to achieve specific goals and tasks through flexible adaptation" (Haenlein, Kaplan, Tan, Zhang, 2019, pp. 341–343). AI is also described as an evolving field of research, encompassing a set of technological capabilities that provide new resources, possibili ties, and applications for organizations and society (Davenport, Guha, Grewal, Bressgott, 2020, pp. 24–42, and Cillo, Rubera, 2024). In some AI-related publications, it is noted that defining AI is challenging since the concept of "intelligence" itself is not mathematically defined (Hutter, 2005, p. V). In the areas of management

DOI: 10.4324/9781003586937-16

and business, individual authors often rely on working definitions of AI. The European Commission has developed a definition of AI in its documents, stating that it refers to

> systems that display intelligent behavior by analyzing their environment and taking actions—with some degree of autonomy—to achieve specific goals. AI-based systems can be purely software-based, acting in the virtual world (e.g., voice assistants, image analysis software, search engines, speech and face recognition systems), or AI can be embedded in hardware devices (e.g., advanced robots, autonomous cars, drones, or Internet of Things applications).
> (Communication from the Commission to the European Parliament, 2018)

Expectations regarding this technology are multifaceted and involve more efficient use of existing knowledge in medical, engineering, and business management fields. AI is seen as an "accelerator" that enhances the practical application of existing technologies, large datasets, and knowledge from various domains—such as medicine, pharmacy, and robotics (Tian, Zhao, Yunfang, Wang, 2023). It is also helpful in gathering data from various departments within a company.

In the literature, many typologies of AI can be found. One of the more interesting ones is the division proposed by Rust and Huang (2021), who identified the following types of AI: (1) mechanical AI, which refers to specific actions of robots; (2) thinking AI, which takes the form of agents capable of interaction and is largely correlated with cognitive functions such as the ability to learn, remember, and solve problems; and (3) feeling AI, which performs routine tasks and can read emotions. In the management literature, another AI typology is highlighted (Davenport, Ronanki, 2018): (1) robotic process automation, which involves the relatively most economically efficient implementation of AI in organizations; (2) cognitive perception, which uses algorithms and machine learning. The goal of this type of AI is to identify specific data features (patterns) and interpret their significance for the organization; and (3) cognitive engagement, which involves using algorithms to process data into natural language for communication with external environments. Collaboration with such intelligent agents (cobots) supports the decision-making process.

The essence of generative AI

The market success in 2022 of OpenAI, which popularized its product ChatGPT as a natural language processing tool based on a large language model, enabled capabilities such as knowledge retrieval, translation across multiple languages, writing code scripts, creating images, and more. Many companies have shown interest in the commercial development of this tool. Various researchers highlight (Utterback, Abernathy, 1975) that "GenAI" is in the "era of ferment," meaning it is in the initial "fluid phase" of development (Mariani, Dwivedi,

2024). In the case of this technology, there is a desire to apply it more broadly in management, while at the same time, there is natural skepticism related to uncertainty about whether this technology could be replaced by something equally valuable. "GenAI" is defined as a type of AI technology designed to generate various types of content, including text, images, sound (Lawton, 2023), and combinations of these elements. The literature (Mariani, Dwivedi, 2024) indicates that "GenAI" is particularly important for creative industries, where it can both initiate innovation and modify the way businesses operate. In the media industry, technologies (Mariani, Dwivedi, 2024) are being sought that could use GenAI to automate many areas of human labor, such as editing articles on sports events, international affairs, financial markets, and more (Trejos-Gil, Gómez-Monsalve, 2024).

The literature emphasizes that the basis for the success of the use of GenAI is mainly the development of a strategy for the implementation of AI by providing data to train the right automatic responses. Among the negative effects of the implementation of AI in the distribution of information by media giants, a long-term decline in advertising revenues in local publishing houses (Abernathy, 2023) in the US is indicated, and consequently "information desertification" consisting in depriving some residents of information. A company's strategy toward the implementation of AI requires an interdisciplinary approach to assessing the risks that must be identified in relation to, among others, cybersecurity, corporate social responsibility, and ethics (Wilczek, Haim, Thurman, 2024, pp. 200–211). In the creative industry, and in particular the film industry, the use of "GenAI" is expected to grow rapidly. It is estimated that in the near future, we will see film productions created mainly with the participation of GenAI. The diffusion of AI in the film industry has an impact on the labor market, affects creative workers, and changes relationships within organizations (Jin, 2024). Changes are also taking place in other industries, thanks to "GenAI" in the chemical industry, creative workers are able to design new chemical units (Yuan, Guo, Wang, 2024) without complicated laboratory procedures, but it is important to have complete data on chemical processes.

Directions of AI research

Despite the numerous studies in the literature, it is emphasized that the extent of the potential opportunities offered by the new technology called "GenAI" is currently unknown. One of the natural applications of this technology is in the area of product innovation, although its use is now more widespread than in the past. It relates to a number of functional areas such as finding optimal suppliers, identifying characteristics of target markets, optimizing production, and marketing activities, among others. It is pointed out (Lawton, 2023) that the publicity of the potential of GenAI is multiplied by the simplicity of the new interfaces, which allow to create the indicated content in a relatively easy way.

When considering the essence of generative AI (Lawton, 2023), it is defined as the ability of algorithms to generate both new content, images and sounds, or programming code. The capabilities of this technology make it possible to automate many functions in a company, both related to data collection and processing, the sphere of production, or marketing activities (Osadchaya Marder, Yule et al., 2024).

Credibility and trust in AI

In the development of AI, efforts are being made to establish ethical boundaries for its activities. Trust in AI and its continuation in the form of GenAI relies on an approach that ensures the systems of machine learning algorithms being developed are responsible and that users have a high degree of trust in them (Ethics Guidelines for Trustworthy AI, 2019). This can be particularly challenging in fields such as medical sciences (Galiana, Gudino, González, 2024, pp. 178–186) due to the sensitivity of the issues involved. The following principles for responsible AI systems are emphasized: (1) respect for human autonomy, (2) prevention of harm, (3) fairness, and (4) the ability to provide explanations. These principles serve as the primary foundation for developing the key requirements that algorithms must meet, which include:

1 Responsibility toward the individual, ensuring actions are taken for their benefit, as well as for the broader community and the natural environment;
2 Human agency, involving a clear distinction between autonomous decisions left to AI and decisions that remain solely within the domain of human judgment;
3 Ensuring a minimum level of reliability in the AI model, such as freedom from racist algorithms or xenophobic data.

Trust in AI hinges on three essential criteria: first, compliance with legal requirements; second, adherence to ethics that safeguard principles and values; and third, the technological performance of the system. These elements collectively contribute to an individual's subjective sense of trust in the information generated by AI and GenAI systems.

Methodology

AI is in the early stages of implementation by various organizations, including enterprises. Many publications take a different view of the nature of AI and its use in enterprises. The research conducted is based on a literature review, where two full-text databases from Elsevier and Springer were used. The literature review made it possible to diagnose the perceptions of the essence of AI, the directions of its application, and allowed to identify critical factors limiting its use. The chapter formulated the following questions: (1) What is AI? (2) How do companies use AI for business purposes? and (3) What barriers can be identified to the use of AI in businesses?

Results

The literature has explored, among other areas, the application of AI in medical (nursing) education, highlighting the need for AI developments such as *AI-assisted simulation robots*, *data mining and machine learning*, and *language models based on natural language processing and deep learning* (Luo, Mao, Wu, He, 2024). AI is seen as a set of algorithms capable of learning from provided data. The essence of generating creative solutions through AI lies, on the one hand, in obtaining data via sensors and detectors (AI perception) and processing them using a set of individual algorithms. On the other hand, AI possesses a system for data processing, "learning," "understanding," and decision-making. AI systems should be identified as a set of integrated inputs from sensor and detector systems, processed according to algorithmic models. Collected external data—such as weather or business information—and internal enterprise data are transformed into actionable information by algorithms responsible for logic and reasoning. These algorithms convert data from databases into information that can be understood by other machines, devices, and/or humans. The information processed and understood by the machine is often presented in simplified form using natural language, intended for human users. Data processing often involves identifying the type of data being processed. Data processing algorithms must (1) interpret the data type and (2) convert the data into a numerical model to make decisions, based on predefined algorithmic systems, that align with the defined goal. All processes aim to develop logical and reliable premises for decision-making, which must be trustworthy for human decision-makers. The concept of "decision" is understood here as the choice of action made by a human, as opposed to a machine, which builds premises for decision-making based on data and algorithms (Figure 13.1). AI systems should not be autonomous in critical areas (such as medical contexts), although they can be autonomous in less critical applications, like regulating traffic lights. A decision can also be a recommendation passed on to a human, who must remain the ultimate decision-maker. In some cases, AI systems may not choose the optimal course of action for a given goal due to data limitations, time constraints, or computational power. A learning rational AI system, on the other hand, identifies and evaluates the state of the business environment (perception) after an action is taken and then adjusts its reasoning rules and decision-making methods.

Under the influence of the development of AI, all areas of social life related to services are changing, in particular education, medicine (Luo, Mao, Wu et al., 2024), architecture, construction (Qin, Guan, Liao et al., 2024) and insurance (Jędrzejczyk, Pizło, 2025, pp. 271–283), as well as all functional areas of marketing research companies (Akter, Sultana, Mariani et al., 2023, pp. 243–261)

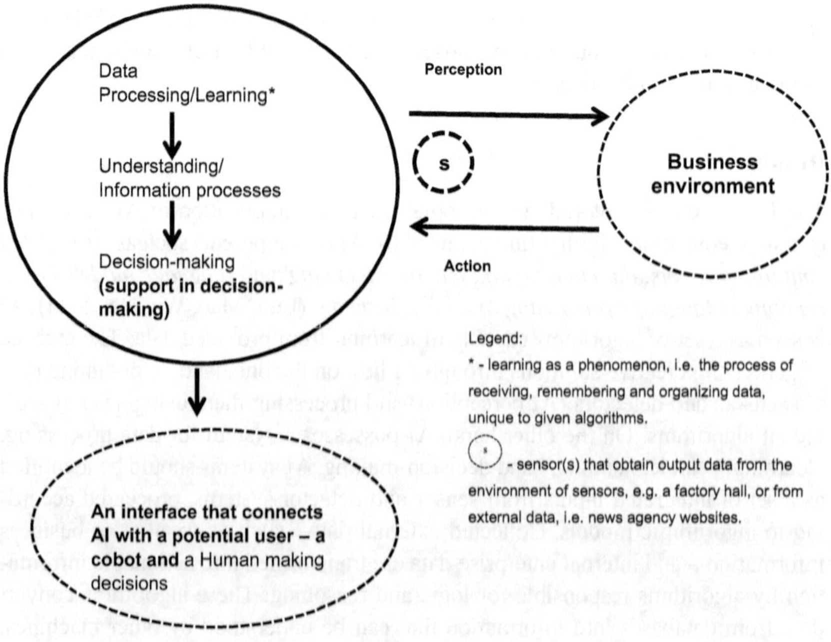

Figure 13.1 Artificial intelligence in the business environment AI—business environment—interface

Source: Own study

and marketing, logistics to finance (Fan, Ilk, Kumar et al., 2024). Qualitative research is usually used to identify research on AI (Jorzik, Antonio, Kanbach et al., 2024) or to combine qualitative research with quantitative research supported by statistical analysis. A significant part of the research conducted on AI focuses on the role (Jorzik, Antonio, Kanbach et al., 2024) (Grashof, Kopka, 2023, pp. 771–797). The implementation of AI in enterprises contributes to improving its productivity, fosters technological progress (Lundvall, Rikap, 2022), and changes the scope of its operation, including the relationship between the employee armed with AI and the company's authorities.

Conclusions

The use of AI will increase in various areas of science and business. The involvement of AI in companies' computer systems will be greater and more precise if their entities use improved algorithms and use high-quality Big Data (BD) to support proper decision-making. More and more tools are being developed that are based on AI. These are various types of translators (Deepl, Google Scholar), editing, and data management programs. Nowadays, AI should be understood as the ability of algorithms operating in computing devices to logically (to the extent of the algorithm's capabilities and the computing power of the machine) interpretation of data and to recognize, adapt, and remember important information based on Big Data in order to achieve goals defined by a given entity. And although, as it is pointed out in the literature, machine learning, marketing called AI, does not meet the definition premise of the concept of intelligence, according to, for example, Hutter (2005), AI is an interdisciplinary area of research that has both general cognitive and scientific applications, as well as purely business and commercial applications. In the first case, this allows for non-laboratory research using temporary knowledge in the field of applied sciences, in the second case, it may refer to the use of AI skills to predict future events and human behavior, for example, in the construction of autonomous and automated production lines and robotic business areas. Enterprises can potentially use AI in all functional areas. They can automate office activities, production processes, including the automation of robotic processes, and, to some extent, marketing activities. The development of AI will progress, and more perfect algorithms will be created to search for information. There will also be a specialization of AI used on the one hand in medicine, insurance, military, and on the other hand, forecasting human behavior in the workplace and cybersecurity of enterprises and critical infrastructure, innovation, and the production process. It should be pointed out that there is an intensive development of machine learning (AI) in technical sciences and in those disciplines of knowledge that have large databases at their disposal, such as medicine and nursing (Luo, Mao, Wu et al., 2024), criminology and the judiciary, and meteorology forecasting weather events.

When identifying the directions of AI development, it is necessary to point out the following features that will be developed in the near future, i.e., the ability to acquire data and the ability to organize it (detecting patterns and certain regularities), logical reasoning, and remembering (deep learning). An important feature of the indicated groups of AI algorithms is the ability to adapt to changing goals, and thus effectively perform a variety of tasks. When formulating the imperfections of AI, one can point to its limitations, failure to see the context of occurring events and regularities, and above all, lack of trust in its results. AI works under strictly defined conditions and its activity is different from solving decision-making problems by humans. The use of AI is associated not only with issues of trust but also with ethics. AI activities often rely on sensitive data—human behavior as an employee (labor market), student (education), and other behaviors remaining in the private sphere. Data processing and formulating conclusions may violate both privacy and suggest actions toward an individual based on questionable decision-making premises. Doubtful, because it is burdened with a high risk of error and the inability to verify the premises of the decision.

References

Abernathy, P. M. (2023). The State of Local News. *Local News Initiative*, November 16, 2023, https://localnewsinitiative.northwestern.edu/projects/state-of-local-news/2023/report/

Akter, S., Sultana, S., Mariani, M., Wamba, S. F., Spanaki, K., Dwivedi, Y. K. (2023). Advancing algorithmic bias management capabilities in AI-driven marketing analytics research, *Industrial Marketing Management,* 114, 243–261, https://doi.org/10.1016/j.indmarman.2023.08.013

Cillo, P., Rubera, G. (2024). Generative AI in innovation and marketing processes: a roadmap of research opportunities, *Journal of the Academy of Marketing Science,* https://doi.org/10.1007/s11747-024-01044-7

Communication from the Commission to the European Parliament, the European Council, the Council, the European Economic and Social Committee and the Committee of the Regions on Artificia lIntelligence for Europe, Brussels, 25.4.2018 COM (2018) 237 final.

Davenport, T., Guha, A., Grewal, D., Bressgott, T. (2020). How artificial intelligence will change the future of marketing, *Journal of the Academy of Marketing Science,* 48(1), 24–42, https://doi.org/10.1007/s11747-019-00696-0

Davenport, T. H., Ronanki, R. (2018). Artificial intelligence for the real world. *Harvard Business Review (HBR)*, https://www.bizjournals.com/boston/news/2018/01/09/hbr-artificial-intelligence-for-the-real-world.html

Ethics Guidelines for Trustworthy AI (2019). *Independent High-Level Expert Group on Artificial Intelligence Set Up the European Commission, European Commission*, B-1049, Brussels, https://digital-strategy.ec.europa.eu/en/library/ethics-guidelines-trustworthy-ai

Fan, S., Ilk, N., Kumar, A., Xu, R., Zhao, J. L. (2024). Blockchain as a trust machine: from disillusionment to enlightenment in the era of generative AI, *Decision Support Systems,* 182, 114251. https://doi.org/10.1016/j.dss.2024.114251

Galiana, L. I., Gudino, L. C., González, P. M. (2024). Ethics and artificial intelligence, *Revista Clínica Española (English Edition)*, 224(3), 178–186, https://doi.org/10.1016/j.rceng.2024.02.003

Grashof, N., Kopka, A. (2023). Artificial intelligence and radical innovation: an opportunity for all companies? *Small Business Economics*, 61, 771–797, https://doi.org/10.1007/s11187-022-00698-3

Haenlein, M., Kaplan, A., Tan, C.-W., Zhang, P. (2019). Artificial intelligence (AI) and management analytics, *Journal of Management Analytics,* 6, 341–343, https://doi.org/10.1080/23270012.2019.1699876

Hutter, M. (2005). *Universal Artificial Intelligence: Sequential Decisions Based on Algorithmic Probability*. Springer-Verlag, Berlin, Heidelberg.

Jędrzejczyk, I., Pizło, W. (2025). The impact of an artificial intelligence on the information asymmetry and the trust of the actors of the insurance market. In *Trust and Artificial Intelligence: Development and Application of AI Technology*, ed. Paliszkiewicz, J., Gołuchowski, J., Routledge, 271–283.

Jin, D. Y. (2024). AI in cultural production in the Korean cultural industries, *Telematics and Informatics Reports,* 13, 100113, https://doi.org/10.1016/j.teler.2023.100113.

Jorzik, Ph., Antonio, J. L., Kanbach, D. K., Kallmuenzer, A., Kraus, S. (2024). Sowing the seeds for sustainability: a business model innovation perspective on artificial intelligence in green technology startups, *Technological Forecasting and Social Change,* 208, 123653, https://doi.org/10.1016/j.techfore.2024.123653

Kaplan, A. M., & Haenlein, M. (2019). Siri, Siri, in my hand: Who's the fairest in the land? On the interpretations, illustrations, and implications of artificial intelligence, *Business Horizons*, 62(1), 15–25, https://doi.org/10.1016/j.bushor.2018.08.004

Lawton, G. (2023). What is generative AI? everything you need to know. *Tech Target.* Available at https://www.techtarget.com/searchenterpriseai/definition/generative-AI

Lundvall, B. Å., Rikap, C. (2022). China's catching-up in artificial intelligence seen as a co-evolution of corporate and national innovation systems, *Research Policy,* 51(1), 123653, https://doi.org/10.1016/j.respol.2021.104395

Luo, Ch., Mao, B., Wu, Y., He, Y. (2024). The research hotspots and theme trends of artificial intelligence in nurse education: a bibliometric analysis from 1994 to 2023, *Nurse Education Today,* 141, 106321, https://doi.org/10.1016/j.nedt.2024.106321

Mariani, M., Dwivedi, Y. K. (2024). Generative artificial intelligence in innovation management: a preview of future research developments, *Journal of Business Research,* 175, https://doi.org/10.1016/j.jbusres.2024.114542

Osadchaya, E., Marder, B., Yule, J. A., Yau, A., Lavertu, L., Stylos, N., Oliver, S., Angell, R., de Regt, A., Gao, L., Qi, K., Zhang, W. Z., Zhang, Y., Li, J., AlRabiah, S. (2024). To ChatGPT, or not to ChatGPT: Navigating the paradoxes of generative AI in the advertising industry, *Business Horizons,* 67(5), 571–581, https://doi.org/10.1016/j.bushor.2024.05.002

Pizło, W. (2009). *Enterprise in the Information Society in the Light of the Theory of Institutional Economics [Przedsiębiorstwo w społeczeństwie informacyjnym w świetle teorii ekonomii instytucjonalnej]*, SGGW, Warszawa, 75–108.

Qin, S., Guan, H., Liao, W., Gu, Y., Zheng, Z., Xue, H., Lu, X. (2024). Intelligent design and optimization system for shear wall structures based on large language models and generative artificial intelligence, *Journal of Building Engineering,* 95, https://doi.org/10.1016/j.jobe.2024.109996.

Rust, R. T., Huang, M. H. (2021). *The Feeling Economy*, Palgrave Macmillan, Cham, 1`11`1q` https://doi.org/10.1007/978-3-030-52977-2_13

Tian, H., Zhao, L., Yunfang, L., Wang, W. (2023). Can enterprise green technology innovation performance achieve "corner overtaking" by using artificial intelligence?—evidence from Chinese manufacturing enterprises, *Technological Forecasting and Social Change,* 194, https://doi.org/10.1016/j.techfore.2023.122732

Trejos-Gil, C. A., Gómez-Monsalve, W. D. (2024). Artificial intelligence in media and journalism: systematic review on Spain and Latin America in Scopus and web of science databases (2018–2022), *Palabra Clave,* 27(4), e2741, https://doi.org/10.5294/pacla.2024.27.4.1

Utterback, J. M., Abernathy, W. J. (1975). A dynamic model of process and product innovation, *Omega,* 3(6), 639–656, https://doi.org/10.1016/0305-0483(75)90068-7

Wilczek, B., Haim, M., Thurman, N. (2024). Transforming the value chain of local journalism with artificial intelligence, *AI Magazine,* 45, 200–211, https://doi.org/10.1002/aaai.12174

Yuan, M., Guo, Q., Wang, Y. (2024). The current research status and prospects of AI in chemical science, *Progress in Natural Science: Materials International,* 34(5), 859–872, https://doi.org/10.1016/j.pnsc.2024.08.003

14 The role of building consumer experience in e-health services

Anna Dąbrowska, Irena Ozimek and Liudmyla Shulhina

Introduction

Experience according to Verhoef et al. (2009) is the sum of feelings resulting from all human-business interactions, occurring at every stage (information seeking, purchase, consumption, etc.) and in every channel...'. Customer experience management is a new paradigm in services, an increasingly important element of marketing strategy, customer satisfaction management (Dąbrowska & Janoś-Kresło, 2019). In recent years, we have witnessed fundamental changes in healthcare, with a surge in the popularity of healthcare virtualisation and digital innovations, to mention telemedicine, artificial intelligence (AI) and mobile applications in healthcare. The aim of the theoretical and empirical chapter is to identify the role of experience in building trust in innovative e-health services. The empirical section will present the results of a survey conducted on representative samples in Poland, followed by Estonia and Spain.

Building the consumer experience against the development of the artificial in the light of the literature

Creating an excellent customer experience is a way to stand out in the market. Customer experience (CX) is defined as the sum of all interactions between the customer and the organisation. It is also defined as the mix of customer service and the emotions it evokes, measured against customer expectations at all touchpoints. In doing so, it is emphasised that for customer experience to be excellent, the entire organisation must work together to deliver an excellent customer experience (Tkaczyk, 2022). Pisano et al. (2015) point out that CX is an element that should determine companies' choices regarding value propositions, distribution channels, customer relationships and revenue streams.

Both CX (Customer Experience) and UX (User Experience) are concerned with building positive feelings on the part of the recipient of a brand's services and helping to address concerns raised on their part, with CX being a broader concept – it is concerned with the overall experience, while UX is concerned

DOI: 10.4324/9781003586937-17

solely with the experience of the digital interface (Tkaczyk, 2022). Furthermore, customer and user experience (CX and UX) are issues of great importance and interest, both among managers and researchers (Szuszkiewicz et al., 2024). The ongoing digital transformation means that more and more products are digital, requiring software development and attention to user experience (User Experi ence, UX). Digital technologies include Big Data, Internet of Things, analytics, social media, cloud computing, AI, blockchain technology, machine learning, Robotic Process Automation, among others (Bartczak, 2023).

According to a Deloitte study, higher consumer demands translate into higher expectations of the channels in which interactions take place (Deloitte, 2023). As highlighted by Szuszkiewicz et al. (2024), digitalisation, reflected in the operation of an increasing number of digital technology platforms, leads to a significant increase in user experience. Personalised service has the greatest impact on customer and employee experience and operational excellence. The coming AI revolution will enable and require the robust use of customer data and insights to offer personalised recommendations for businesses to understand customers and increase value. GenAI will now enable businesses to provide their customers with a much better experience when searching the knowledge base to solve problems or create products (Deloitte, 2023).

When talking about the importance of building the consumer experience especially in healthcare, it is also impossible not to mention AI, which is having an increasing impact on the development of this sector and on the services provided to consumer-patients. It should be emphasised that health is one of the basic human values.

Over the past few decades, AI in medicine has opened up new possibilities in the diagnosis, treatment and care of patients, but central to this process is the building of trust among both doctors and patients, which influences the positive consumer experience of these services.

The dynamics of AI development prove that it is becoming an integral part of modern global medicine. It undoubtedly contributes to the introduction of advances in healthcare and thus more effective treatment of patients, which is very important in view of the expanding catalogue of diseases of civilisation. (Kovačovicová & Vaňhara, 2023). The application of AI covers various areas of medicine, such as cardiovascular medicine (Siontis et al., 2021), complex lung diseases (Exarchos et al., 2022), in dentistry (Tomášik et al., 2024), early detection and prevention of oral diseases, or in imaging (Kamath et al., 2024).

The positive experience of using AI will be influenced by the benefits of AI. Not only are improvements in the diagnostic process, personalisation of treatment, therapeutic decision support, innovations in robotic surgery and patient monitoring pointed to, but medical education, healthcare management and mental health are also mentioned (AIoAI, 2023).

The study *What doctor? Why AI and robotics will define New Health* shows that as many as 55% of patients are ready for doctors to be replaced by high-tech and AI-equipped robots that could answer health questions, perform tests, make diagnoses and recommend treatments (PwC Report, 2017). This may indicate a high level of confidence in AI. More than 11,000 people from 12 countries in Europe, the Middle East and Africa took part in this survey. Emerging markets are the most open to the use of new technologies in healthcare, hence countries such as Nigeria (94%), Turkey (85%) and South Africa (82%) led the way. In Western European countries, indications were lower: the UK (39%), Germany (41%), Sweden (48%) and Norway (50%), but suggesting interest in such solutions. In addition, the PwC survey found that almost half, up to 73%, of respondents would prefer minor surgical procedures to be performed by robots rather than doctors. This situation did not apply to major surgical procedures such as knee or hip replacements, tumour removal or heart surgery. Nevertheless, from 27% in the UK, to 40% in the Netherlands, to 69% in Nigeria, respondents were willing to undergo a major procedure performed by a robot.

However, every revolutionary change brings not only certain benefits, but also certain risks and threats. Among these are: diagnostic errors, privacy, lack of control, ethical dilemmas, exclusion, security, costs, data dependency, the human element (possibility of loss of personal contact between doctor and patient) (Musioł, 2023). Despite various concerns, the development of healthcare without AI is impossible.

It is worth mentioning that on 13 March 2024, the European Parliament approved the Artificial Intelligence Act, (Artificial Intelligence Act), which ensures security and respect for fundamental rights while promoting innovation. The aim of the new legislation is to foster trustworthy AI in Europe and beyond by ensuring that AI systems comply with fundamental rights, security and ethical principles, and by countering the threats posed by very powerful and influential AI models (European Parliament, 2024). This is very important in view of the fact that the true footprint of AI in healthcare is still limited and is only at the beginning of its full potential.

Objective and methodology of the study

The aim of the survey was to identify consumers' experience of e-health services and their confidence in innovative solutions.

To achieve the aim of the study, it was necessary to seek answers to the following research questions:

- whether respondents would benefit from innovative health e-services,
- do they have confidence in health e-services?
- how do they assess teleportation?

The survey among Poles was conducted on 30.11–12.12.2021 on a sample of 1,001 respondents aged 18–65+. It was a representative sample of the adult population of Poland in terms of the total distribution of gender, age category and size class of place of residence, which allowed the results obtained and conclusions formulated to be generalised to the entire population of Poles. The research was conducted using the data collection technique: CAWI – online interviews on the SW PANEL research panel belonging to the SW Research agency, which carried out the survey. The sample selection was random. The research tool was a survey questionnaire, which consisted of 29 basic questions that corresponded to the adopted research questions and four metric questions (age, gender, respondent's education, place of residence). The questions answered the research questions posed.

The profile of the average respondent was as follows: female (53.2%), aged 26–35 or 56–65 (20.6% each), living in rural areas and towns with up to 9,000 inhabitants (40.6%), with a university degree (44.8%).

The study was carried out as part of the MEiN grant for the maintenance and development of research potential at the College of Management and Finance at the Warsaw School of Economics in 2021 awarded for the project "Consumer behaviour in the e-health market during a pandemic" (study leader: Anna Dąbrowska).

The international survey covered two countries, Estonia and Spain. It was conducted in November/December 2023. The survey covered 528 Estonians and 526 Spaniards aged 18–65+. The survey instrument was a questionnaire consisting of 15 factual questions and four metric questions (gender, age, education level, nature and size of place of residence). The survey was carried out using the CAWI method, by the research company ARC Rynek i Opinia using the resident bases in the respective survey panels in each country. Participation in the survey was voluntary and anonymous.

The profile of the average respondent in Estonia is: female (52.6%), aged 35–44 (27.8%), living in a city with up to 49,000 inhabitants (32.4%), with post-secondary, secondary, basic vocational education (49.1%). The profile of the average Spanish respondent, on the other hand, is male (59.3%), aged 35–44 (30.0%), living in a city with 500,000 or more inhabitants (27.2%), with a university degree (53.2%).

The study was carried out as part of the 2,023 collegiate staff development grant at the College of Management and Finance at the Warsaw School of Economics awarded for the project "Consumer behaviour in the e-health market in Estonia and Spain" (study leader: Prof. Anna Dąbrowska, PhD).

In both surveys, a 5-point Likert-type positional scale was used to measure respondents' opinions, attitudes and views (e.g. on the importance, usefulness or threats of health e-service tools, e.g. 1 – definitely not, 2 – rather not, 3 – hard to say, 4 – rather yes, 5 – definitely yes, and in the case of evaluating the importance of e-services: 1 – definitely not important/not useful, 2 – rather not important, 3 – hard to say, 4 – rather important, 5 – definitely important.

In the case of the survey conducted among Poles, as well as the survey conducted among Estonians and Spaniards, people who had used at least two e-health services in the last 12 months prior to the questionnaire survey were eligible for the study. Two demographic characteristics were taken into account when discussing the survey results: gender and age.

Research findings

Both studies sought to elicit responses from respondents to the research questions:

- Would respondents benefit from e-health technological innovation?
- do they have confidence in e-health services?
- how do they assess teleportation?

Before moving on to answer the research question of whether Polish respondents would benefit from e-health technological innovations, it is important to note that a government pilot programme testing the usefulness of electronic stethoscopes for remote patient examinations at home began in Poland in July 2021. An e-stethoscope is an instrument that converts sound into electrical signals, which can then be amplified, sent over the internet as a file, or subjected to automated analysis. The recording from the e-stethoscope can also be listened to in real time and evaluated by a doctor. The Ministry of Health has earmarked approximately PLN 5.5 million for the purchase of stethoscopes and other activities under the programme (ForumeZdrowia, 2021).

In the case of the survey among Polish respondents, negative answers (definitely no and rather no) outweighed positive answers (definitely yes and rather yes). Noteworthy are the neutral answers (difficult to say), which ranged from 22.9% to 32.3% (Figure 14.1).

Despite the great facilitation in controlling the state of their health, Polish respondents showed a rather high degree of restraint. It was noted that the highest propensity to use innovative e-health services was declared by consumers with regard to remote diabetes testing (31.6% of responses), which would definitely

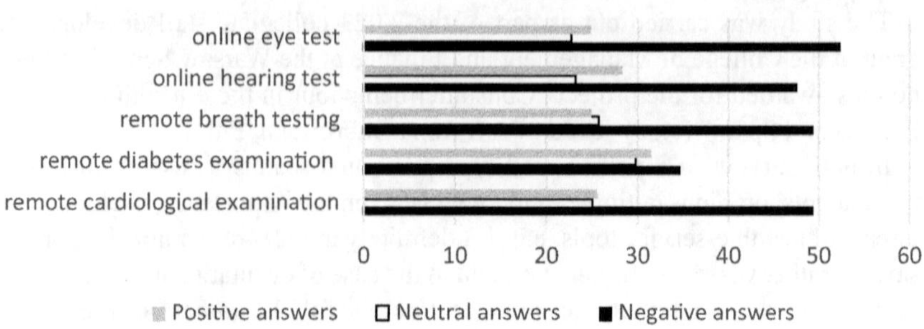

Figure 14.1 Willingness of Poles to use innovative e-health services (% of responses)

Source: Own survey, 2021

Table 14.1 Willingness of Estonians and Spaniards to use innovative e-health services (% of responses)

Innovative e-health services	*Negative answers*		*Neutral answers*		*Positive answers*	
	Estonia	*Spain*	*Estonia*	*Spain*	*Estonia*	*Spain*
Remote cardiological examination	44.2	45.8	11.3	3.2	44.5	50.9
Remote diabetes examination	39.8	36.7	14.4	4.8	45.8	58.6
Remote testing Respiratory	46.8	45.2	11.2	3.4	42.0	51.3
Online hearing tests	40.3	36.9	11.0	2.1	48.7	61.0
Online eye tests	40.0	42.2	10.8	2.1	49.2	55.7

Source: Own survey, 2023

be used by: 30.6% of women, 31.8% of men. For age, the following correlations were noted: 26.1% of those aged 18–25, 34.5% of those aged 26–35, 28.8% of those aged 36–45; 36.5% of those aged 46–54, 30.5% of those aged 55–64 and 32.0% of those aged 66–80.

In contrast, 28.3% of respondents, including 23.5% of women and 33.5% of men, said they were willing to use an online hearing test. Among the different age groups, the distribution of responses was as follows: 32.3% of those aged 18–25, 29.1% of those aged 26–35, 26.5% of those aged 36–45; 31.8% of those aged 46–54, 24.7% of those aged 55–64 and 25.2% of those aged 66–80.

As mentioned, a survey on eHealth services in Estonia and Spain was conducted in 2023. Respondents were asked about their willingness to use innovative eHealth services as part of the e-stethoscope – Table 14.1.

Analysing the results of the survey, it can be said that positive answers predominate, which may indicate the respondents' favourable experience and their openness to and confidence in innovative solutions in health e-services. In Spain, there were small percentages of "hard to say" responses, which can be interpreted as having knowledge in the area surveyed. The response rates are more favourable than for Polish respondents.

In Estonia, the highest positive indications were for online eye examination (49.2% of responses) and online hearing examination (48.7%). There were slightly fewer women (45.2%) than men (54.3%) in favour of online eye examination. By age group, the distribution of responses was as follows: 50.0% of those aged 18–24, 42.3% of those aged 25–34, 51.0% of those aged 35–44, 50.4% of those aged 45–54 and 48.7% of those aged 55–80. In contrast, men (53.4%) were slightly more likely to be willing to test their hearing online than women (44.9%). Considering the age of the survey participants, 50.0% of 18–24 year olds, 42.3% of 25–34 year olds, 51% of 35–44 year olds, 50.4% of 45–54 year olds and 48.7% of 55–80 year olds declared using this type of test.

In contrast, Hispanic respondents mainly indicated an online hearing test (61.0% of responses). Men (65.0%) were more likely to declare this type of survey than women (55.2%). The distribution of responses by age of survey participants for this type of survey was as follows: 66.7% of those aged 18–24, 65.7% of those aged 25–34, 58.2% of those aged 35–44, 62.1% of those aged 45–54 and 56.1% of those aged 55–80.

Remote diabetes testing was indicated by 58.6% of Spanish respondents, including: 55.6% of women and 59.9% of men. Considering the age of the respondents, the responses by age group were as follows: 60.4% of those aged 18–24, 53.3% of those aged 25–34, 57.0% of those aged 35–44, 64.6% of those aged 45–54 and 58.3% of those aged 55–80.

The results indicate that consumers, depending on their country of origin, showed very different interest in the various forms of online health services. Thus, Polish survey participants were most interested in remote diabetes testing and least interested in online eye examinations. Respondents from Spain, on the other hand, were most interested in online hearing tests and least interested in online cardiology tests. Estonian participants, on the other hand, expressed the greatest interest in online eye examination and the least interest in remote respiratory examination.

The survey also aimed to identify whether e-health services are useful to respondents, whether they trust them. In the case of survey participants with a Polish background, the question of whether they trust e-health services received 30.2% positive answers, 43.3% were neutral answers (neither yes nor no), while 26.7% were negative answers. Estonians, like Spaniards, rated these services very highly, with 86.9% of respondents each giving positive ratings. Negative ratings accounted for, respectively, 9.3% and 12.0%, and neutral (hard to say): 3.8% and 1.1%.

Another context studied concerned telemedicine advice. In the countries surveyed, respondents were asked how they evaluated telemedicine services. In the case of Polish respondents, 56.5% of the respondents declared that they had used telemedicine advice during the COVID-19 pandemic. This was mainly by telephone. However, 31.8% of respondents indicated that they preferred direct contact with a doctor. More than 78% of Estonian respondents and almost 84% of Spanish respondents rated online consultations with a doctor positively. Negative evaluations accounted for, respectively, 16.9% and 13.3%.

Respondents in Estonia and Spain were also asked how they perceive the use of eHealth services in the future. Almost 40% of Estonian respondents said that they intend to increase their use of eHealth services, indicating confidence and a good experience to date, and 56.1% that the level of use will not change. In contrast, almost 50% of Spanish respondents said that they intend to increase their use of e-health services and 47.5% that their level of use will remain the same.

Discussion of the findings

As highlighted in the WHO report, telemedicine is an open and constantly evolving science, as it embraces new technological developments and responds and adapts to the changing health needs and contexts of societies and defines telemedicine as "the delivery of health services by professionals, where distance is a key factor, using communication technologies to exchange relevant information for diagnosis, treatment, prevention, research, consultation or medical knowledge to improve patient health" (WHO Global Observatory for eHealth, 2010). The COVID-19 pandemic has forced individuals and health systems to review what is possible and desirable and to adapt care models to the rapidly changing situation. In many countries, there has been a shift towards telephone and video consultations. In addition, pressure to adopt telehealth solutions to support self-management is currently growing, mainly due to the increasing number of patients with chronic diseases who are living to old age through the pace of medical innovation (Blandford et al., 2020). It is noteworthy in this context that both 40% of Estonian respondents and 50% of Spanish survey participants declared that they intend to increase their use of e-health services. In contrast, an overwhelming majority of Polish respondents (59.3%) believe that digital services can allow citizens to increase control over the management of their own health.

It is important to note a very important context related to telemedicine, namely the potential for risks such as an exacerbation of the digital divide, poor software engineering and security breaches (Blandford et al., 2020). In the self-report survey, the least confidence in healthcare e-services was recorded for survey participants with a Polish background (only 30.2% indicated a positive response), and as many as 43.3% were neutral responses (neither yes nor no). Thus, less than one in three people perceived positive solutions in innovative health services, which may indicate positive experiences they have encountered or assume that innovative services will be a beneficial solution for them. Both Estonians and Spaniards rated them significantly higher (86.9% of respondents gave positive ratings). As highlighted by Galiero et al. (2020), future telemedicine platforms must be secure, reliable and flexible enough to meet the requirements of regulatory, professional and healthcare organisations. All these platforms should be regularly updated. To realise the full potential of telehealth, patients and professionals must trust digital systems to ensure the privacy and security of health information. The latest security and data encryption systems should be implemented to protect patient privacy. Patients need to be aware of the privacy choices they make to ensure data security and avoid disclosure of sensitive personal information. In determining data sharing policies, the interests of the individual must be balanced with those of the population, recognising that different cultures tend to value these interests differently (Blandford et al., 2020).

Conclusion

The development of eHealth services will be determined, among other things, by patient experience and broader safety. to these services. As the results show, respondents declared that they intend to increase their use of eHealth services, indicating trust and good experiences to date. Furthermore, the analysis of the results shows that cultural differences influence the level of perception of eHealth services. Understanding these differences can help to better manage risks and formulate effective strategies tailored to the specific needs and concerns of consumers in using such services.

Future regular surveys in this area should enable a better understanding of regional and demographic differences in consumer perceptions of health-related e-services. In addition, they would enable a better identification of consumer needs and expectations regarding health e-services. Further research taking these trends into account may provide a more comprehensive picture, both of the level of consumer concerns and of further directions for such services.

AI will play a role in the development of innovation in health services. The European Parliament is taking action to ensure safety and respect for fundamental rights while supporting innovation.

It is undeniable that the level of use of health e-services will depend on consumer-patient education. Raising awareness of the positive aspects of innovative solutions takes time and knowledge. Many actors, not only those related to healthcare, must be involved in this process. Education conducted at different levels of science can be of great importance. Awareness of patient rights cannot be absent from this process.

References

AIo AI (2023). *Jak sztuczna inteligencja zmienia świat medycyny: 10 przykładów zastosowań i korzyści, [How artificial intelligence is changing the world of medicine: 10 examples of applications and benefits].* Retrieved April 23, 2024, from https://aioai.pl/jak-sztuczna-inteligencja-zmienia-swiat-medycyny-10-przykladow-zastosowan-i-korzysci/#Zmniejszenie%20błędów%20Medycznyc

Bartczak, K. (2023). *Modele biznesu oparte na cyfrowych platformach technologicznych. [Business models based on digital technology platforms].* Warszawa: Difin.

Blandford, A., Wesson, J., Amalberti, R., AlHzme, R. & Allwihan, R. (2020). Opportunities and challenges form telehealth within, and beyond, a pandemic. *The Lancet Global Health*, *8*(11), November 1. https://doi.org/10.1016/S2214-109X(20)30362-4

Dąbrowska, A. & Janoś-Kresło, M. (2019). The importance of customer experience for service enterprises. *Marketing i Rynek [Journal of Marketing and Market Studies]*, *9*, 15–25.

Deloitte (2023). *Report: customer service excellence 2023.* Retrieved September 22, 2024, from https://www2.deloitte.com/pl/pl/pages/deloitte-digital/Articles/Raport-Customer-Service-Excellence-2023.html

European Parliament (2024). *EU AI Act: first regulation on artificial intelligence. The use of artificial intelligence in the EU will be regulated by the AI Act, the world's first comprehensive AI law. Find out how it will protect you.* Retrieved June 23, 2024, from https://www.europarl.europa.eu/topics/en/article/20230601STO93804/eu-ai-act-first-regulation-on-artificial-intelligence

Exarchos, K., Aggelopoulou, A., Oikonomou, A., Biniskou, T., Beli, V., Antoniadou, E. & Kostikas, K. (2022). Review of artificial intelligence techniques in chronic obstructive lung disease. *IEEE Journal of Biomedical and Health Informatics*, 26(5), 2331–2338.

ForumeZdrowia (2021). *Ruszył program pilotażowy e-stetoskopów. [E-stethoscope pilot programme launched].* Retrieved June 20, 2024, from https://forumezdrowia.pl/info/aktualnosci/ruszyl-program-pilotazowy-e-stetoskopow/

Galiero, R., Pafundi, P., Nevola, R., Rinaldi, L., Acierno, C., Caturano, A., Salvatore, T., Adinolfi, L., Costagliola, C. & Sasso, F.C. (2020). The importance of telemedicine during COVID-19 pandemic: a focus on diabetic retinopathy. https://doi.org/10.1155/2020/9036847

Kamath, P., Kamath, P., Saldanha, S., Shetty, T., Rodrigues, S. J., Mahesh, M., Pai, U., Hegde, P., Rodrigues, S. J. & Mukherjee, S. (2024). A brief exploration of artificial intelligence in dental healthcare: a narrative review [version 2; peer review: 1 approved, 1 approved with reservations]. *F1000Research*, 13(37). https://doi.org/10.12688/f1000research.140481.2

Kovačovicová, P. & Vaňhara, P. (2023). *AI in cancer research: from histology to personalized medicine*. BIOCEV Regeneration III. Retrieved May 5, 2024, from www.muni.cz/vyzkum/publikace/2342320

Musioł, M. (2023). *Sztuczna inteligencja na usługach medycyny. [Artificial intelligence at the service of medicine].* Retrieved June 5, 2024, from https://naturazdrowie.com/2023/09/11/sztuczna-inteligencja-na-uslugach-medycyny-mozliwosci-i-zagrozenia/

Pisano, P., Pironti, M. & Rieple, A. (2015). Identify innovative business models: can innovative business models enable players to react to ongoing or unpredictable trends? *Entrepreneurship Research Journal*, 5(3), 181–199.

PwC Report (2017). *Nowe technologie rewolucjonizują rynek opieki zdrowotnej. [New technologies are revolutionising the healthcare market].* Retrieved February 10, 2024, from pwc.co.uk.

Siontis, K. C., Noseworthy, P. A., Attia, Z. I. & Friedman, P. A. (2021). Artificial intelligence-enhanced electrocardiography in cardiovascular disease management. *Nature Reviews Cardiology*, 18(7), 465–478. https://doi.org/10.1038/s41569-020-00503-2

Szuszkiewicz, A., Twardzik, M. & Lubowiecki-Vikuk, A. (2024). Doświadczenie klienta i użytkownika elementem innowacyjnych modeli biznesowych – perspektywa start-upów. [Customer and user experience in innovative business models: a start-up perspective]. *Studia i Prace Kolegium Zarządzania i Finansów [Studies and Work of the Collegium of Management and Finance],* 196, 107–122. https://doi.org/10.33119/SIP.2024.196.8

Tkaczyk, J. (2022). *Czym jest doświadczenie klienta i dlaczego jest ważne? [What is customer experience and why is it important?].* Retrieved April 23, 2024, from https://www.kozminski.edu.pl/pl/review/czym-jest-doswiadczenie-klienta-i-dlaczego-jest-wazne

Tomášik, J., Zsoldos, M., Oravcová, L., Lifková, M., Pavleová, G., Strunga, M. & Thurzo, A. (2024). AI and face-driven orthodontics: a scoping review of digital advances in diagnosis and treatment planning. *AI,* 5, 158–176. https://doi.org/10.3390/ai5010009

Verhoef, P. C., Lemon, K. N., Parasuraman, A., Roggeveen, A., Tsiros, M. & Schlesinger, L. A. (2009). Customer experience creation: determinants, dynamics and management strategies. *Journal of Retailing, 85*(1), 31–41. https://doi.org/10.1016/j.jretai.2008.11.001

WHO Global Observatory for eHealth (2010). *Telemedicine: opportunities and developments in Member States: report on the second global survey on eHealth.* World Health Organization. Retrieved April 23, 2024, from https://iris.who.int/handle/10665/44497

15 Trust and transparency as facilitators of GenAI-driven transformation of businesses

Knowledge management perspective

Alicja Fandrejewska

Introduction

The advancement and exponential growth of information and communication technologies, including artificial intelligence (AI), machine learning (ML), and natural language processing (NLP), have significantly transformed the business landscape. The processes of digital transformation and the development of new technologies present both opportunities and challenges for organizations, and they significantly impact how they operate. Applying mobile solutions, AI, ML, and NLP, big data, cloud computing, blockchain, and the Internet of Things enable businesses to considerably improve their processes and performance, and innovative IT projects focus on enhanced customer experience, streamlining operations, automating tasks, or even creating new business models.

Changes related to digital transformation, including generative AI-driven solutions, are not limited to large organizations with capital, capacity, and resources enabling them to invest in technologies and innovations (Lokuge & Duan, 2021). Smaller businesses, despite their limited resources, infrastructure, and scope of operations, increasingly realize that in order to stay competitive, they also need to leverage new technological solutions and learn how to use them effectively to improve their performance and financial results. Also, they frequently have the advantage of being agile, flexible, and more responsive to market needs, and they can benefit from it often to a greater extent than large-scale corporations (Lokuge & Duan, 2023; Soni, 2023).

New technologies, including GenAI, can support and shape the creation of enhanced products and improved organizational processes. Entrepreneurs also integrate innovative solutions into existing business processes, organizational and corporate structures, or working models (Warner & Wäger, 2019). However, it is important to bear in mind that at the heart of successful digital transformation lies the ability of leaders to incorporate technology in line with company's strategy and goals and the willingness of employees and managers to effectively

DOI: 10.4324/9781003586937-18

share knowledge and collaborate, with both of these processes being supported by transparency and trust in the employer-employee relationship.

The prerequisites for digital transformation

Digital transformation relies heavily on knowledge management and knowledge-sharing processes in organizations. Knowledge management is a concept closely related to the present stage of development of our economy and society, i.e. digital transformation, information society, or knowledge-based economy (Wiśniewski, 2019, p. 9). Knowledge management activities include knowledge acquisition, encoding, storage, transfer, application, and sharing (Deng & Lu, 2022). Thus, knowledge sharing, covered in this chapter, is an important element of knowledge management and a key factor in improving organizations' strategic, innovative, and marketing abilities (Chatterjee et al., 2022; Yeboah, 2023). Combined with knowledge application, knowledge sharing creates value for the business, and it is crucial for organizational process and performance outcomes (Ahmad & Karim, 2019).

Proper knowledge management and knowledge-sharing culture are essential for the successful implementation of AI-driven solutions in business, and trust is regarded as "a crucial construct and predictor of the willingness to adopt AI systems, garnering increasing attention from researchers, developers, and policymakers" (Riedl, 2022; Kreps et al., 2023 after: Paliszkiewicz & Gołuchowski, 2025). As Paliszkiewicz and Gołuchowski stress, trust in AI is considered not just in terms of its reliability, i.e., accuracy and efficiency of algorithms. It is also perceived through the prism of the ethical implications of its implementation and frequently discussed in the context of "transparency, accountability, and the alignment of AI systems with human values, privacy concerns, and societal norms" (Paliszkiewicz & Gołuchowski, 2025). In this chapter, trust and transparency will be considered primarily from the perspective of the employee-employer relationship.

The role of data, information, and knowledge in digital transformation

The hyper-connected world and its VUCA (volatile, uncertain, complex, and ambiguous) character stress the need to process information effectively to make it useful for action (Bennis & Nanus, 1986; Barber, 1992; Argyris, 1985). Data, information, and knowledge play a key role and are regarded as sources of competitive advantage, especially in the volatile environment in which organizations must operate. As Sayyadi and Provitera (2023) stress, "executives have to deal with causal ambiguity on a daily basis, and it has been defined in multiple interpretations". Facing many challenges related to maintaining their position in the market, businesses started to focus on their capability to create and utilize knowledge as key factors in obtaining a sustainable competitive advantage (Zheng et al., 2010). They also realize that knowledge is the asset they could leverage more efficiently (Liebeskind, 1996), especially with the increased computing power and capabilities expanded with AI, ML, or artificial neural networks. In the digital transformation era, data, information, and knowledge have become "the lifeblood of organizations", driving innovation, efficiency, and competitive advantage (Gul et al., 2020; Zhang & Jin, 2023). Many businesses wish to capitalize on the new AI-driven developments, and advanced data analytics and reliance on information have become common characteristics of companies today.

Knowledge-sharing in the process of digital transformation

As mentioned above, digital transformation relies heavily on high-quality data, information, and knowledge in its many forms and dimensions, as well as intra- and inter-company communication and collaboration. The chances of successfully introducing innovative technological solutions without relying on available knowledge resources and staff members' and managers' know-how, skills, and expertise are scarce, if not non-existent. However, employees are frequently reluctant to transfer or share knowledge, especially in a changing, unstable, highly competitive environment. Operating in conditions of uncertainty and constant change, characteristic of digital transformation, can undermine trust between employees and their managers, which can be detrimental to knowledge acquisition, transfer, and sharing (Lokuge & Duan, 2023; Lazić & Jović, 2019). Also, the present stage of development of AI, including Gen AI, is "a contested concept that lacks clear consensus" in areas such as computer science, law as well as public perception (Larsson & Heintz, 2020), which makes the transition even more challenging.

Companies are eager to apply advanced analytical tools in this changing environment and benefit from high-quality data and actionable information to make better decisions. However, supporting managerial decision-making relies on providing integrated, high-quality information that is properly structured, easily interpreted, and delivered on time. (Hayen et al., 2007). Therefore, data governance and curation procedures must be implemented within the organizations to ensure core data quality dimensions, including completeness, uniqueness, timeliness, validity, and accuracy (*The Six Primary Dimensions for Data Quality Assessment*). Experiencing volatility on a daily basis, organizations cannot rely on long-term forecasts and need to revise their strategy and tactics continuously. Company resources, agility, and responsiveness to market changes are highly valued in this context (Królik, 2015, p. 153). As the author states, agility, and responsiveness stem from the knowledge and experience the organization and individual members have acquired while conducting business and solving current problems. Thus, the knowledge the company has gained presents great potential of an enterprise, which, properly put to use can help to "optimally combine all the attributes of an organization" (Królik, 2015, p. 153).

The relationship between knowledge sharing and trust

The relationship between knowledge-sharing and trust has been the focus of attention of many academics, researchers, and business practitioners (Paliszkiewicz, 2010; Hasnain, 2017; Bulińska-Stangrecka, 2018). Many researchers focused on the relationship between trust and knowledge-sharing. Gilmour et al. (2015) claim that trust of members of an organization is the basis for knowledge exchange. Rolland and Chauvel (2000) claim that trust is "the single most important precondition for knowledge exchange" (p. 239). Davenport and Prusak (1998) find that trust positively influences inter-organizational knowledge transfer. Usoro et al. (2007) focused on the importance of trust in knowledge transfer and knowledge-sharing within virtual communities, considering three dimensions of trust, i.e. competence, integrity, and benevolence. Lucas (2005) examined the impact of trust and reputation on the transfer of best practices, and his study found that trust and reputation have a significant and positive impact on knowledge transfer.

Researchers have analyzed many aspects and dimensions of the relationship between trust and the willingness to share individual expert and tacit knowledge, i.e. "the knowledge created in the human mind as individual know-how and can be expressed as innovation" (Mohajan, 2016). Brilman (2002) believes that trust impacts the process of knowledge creation and sharing. Ashleigh et al. (2003 after: Hasnain, 2017) used the context of communities of practice to establish a model linking knowledge transfer and trust, among others, in knowledge, expertise, skills, capacity, and values. Singh and Premarajan (2007) attempted to assess the effect of trust and culture on tacit and explicit knowledge transfer. In her study, Paliszkiewicz (2007) discovered that insufficient trust levels adversely impact knowledge-sharing processes in SMEs. The results of the study carried out by Holste and Fields (2010) suggested that adequate levels of affect-based and cognitive-based trust are required for managers and professionals to be willing to share and use tacit knowledge. Kucharska (2017) found that "there is a strong correlation between trust and collaborative culture" and "both of them have a strong influence on tacit knowledge sharing".

Other authors focused on the processes and mechanisms in organizations which were supported by knowledge-sharing initiatives and mechanisms in innovative and technology-driven organizations. For instance, Sreckovic and Windsperger (2012) examined the impact of trust on selecting the knowledge transfer mechanisms and found that experience-based trust increases knowledge sharing between the cluster partners by increasing the face-to-face knowledge transfer mechanisms. Sankowska (2013) presented empirical evidence on the relationship between organizational trust, knowledge transfer, creation, and innovativeness at the firm level. The link between the importance of trust, knowledge sharing, and technology-based transformation has been the research focus for many authors and business practitioners. Numerous studies confirmed the crucial importance of trust when dealing with uncertainty and risk related to implementing new technologies in organizations (Lokuge & Duan, 2023; Vial, 2019; Stonehouse & Konina,2020; Trenerry et al., 2021).

The relationship between knowledge sharing and transparency

As mentioned earlier, knowledge sharing is critical to knowledge management and significantly impacts organizational performance. It refers to the voluntary "acts of making knowledge available to others within the organization" (Ipe, 2003, p. 32), which can lead to improved team and organizational productivity, innovation, and competitive advantages (Xia & Yang, 2020). However, a strategic approach to data management, information processing, and knowledge extraction is required for companies to make more informed decisions, stay afloat, and ahead of their competitors.

The research shows that the transparency of leaders, understood as authentic, transparent, and relational leadership, can contribute to the willingness and engagement of staff in knowledge-sharing and facilitate the development of a knowledge-sharing culture (Yeboah, 2023). Ethical leadership, which involves personal ethics and integrity, consistency, transparent communication, and adherence to ethical standards, can support knowledge-sharing processes and build knowledge-sharing culture (Lee & Cheng, 2010). Authentic and transparent leadership also refers to openness and truthfulness regarding the "free exchange of knowledge and information" (Gardner et al., 2005 after: Kempster et al., 2019).

The transparency of the leader may help foster reciprocity in the employer-employee relationship. The exchange of knowledge and information and two-dimensional feedback can contribute to more effective collaboration and higher success rates in IT projects, particularly related to Generative AI. Responsible leadership, based on the principles of social exchange, can also positively influence knowledge sharing and job performance among knowledge workers (Lin et al., 2020). Additionally, relational transparency, accountability of leaders and their "fidelity to purpose" (Kempster et al., 2019) may encourage employees to engage in Gen-AI-driven business transformation and identify with the values of a knowledge-sharing culture.

Challenges resulting from implementing GenAI-based solutions in the workplace

Integrating new technologies, including GenAI, into the workplace has significantly transformed the dynamics within the company and affected relationships between managers and employees. As organizations increasingly embrace digital transformation, there is a growing need to address the issues that arise from the increased use and reliance on AI. Employees engage in new forms of work that involve less personal interaction and more engagement with information technology. This shift can impact their daily tasks, their emotional well-being and create uncertainty about their role within the organization (Mirbabaie et al., 2021; Giuntella et al., 2023). As Mirbabaie et al. state, "introducing AI may negatively impact employees' identifications with their jobs as AI is expected to fundamentally change workplaces and professions, feeding into individuals' fears of being replaced".

Integrating AI-driven technologies can affect the relational aspects of psychological contracts and diminish employee engagement (Braganza et al., 2021). A psychological contract between managers and staff, understood as "an implicit understanding of the relationship between employee and employer" (Coyle-Shapiro & Parzefall, 2008), i.e. the human side of the relationship between an employee and an employer (https://www.aihr.com/blog/psychological-contract/), is affected by the implementation of the AI. This "implicit agreement between an employer and an employee that outlines the unwritten expectations, beliefs, and obligations that both parties have towards each other" (https://www.sloneek.com/lexicon/psychological-contract/) needs to be rectified at the time of GenAI-enabled digital transformation since the role, the scope of obligations and basic assumptions, change due to new technological developments.

There are specific conditions which need to be met for the digital transformation to be successful. Dombrowski et al. (2007) indicate that it requires building and maintaining a specific innovative organizational culture with creativity and adaptability seen as priorities; communication channels encouraging open dialogue and free flow of ideas; safe spaces which allow for risk-taking, experimentation, and exploration of new approaches; flexible processes and structures accommodating changing needs and new opportunities; flexible boundaries allowing the organization to connect with external sources of knowledge and innovation; incentive schemes recognizing contributions and authentic and transparent leadership.

It is important to note that the rapid advancements in AI have transformed various sectors and industries, reshaping how companies operate and the skills required from their workforce. AI has become a disruptive force, incorporated into core business processes and impacting employees, various occupations, and sectors (George, 2024). As AI continues to evolve, it is crucial to understand its multifaceted influence on the workplace and the challenges it presents for employees and managers. As AI becomes more prevalent in the workplace, employees need to develop new skills and competencies to remain relevant and competitive, and organizations must invest in reskilling and upskilling initiatives (*How AI Will Transform…*). While AI can potentially displace certain jobs, it also creates new opportunities for human-machine collaboration. AI can enhance human capabilities, allowing employees to focus on more complex, creative, and strategic tasks.

Another concern related to AI implementation is infringing or violating intellectual property rights. Employees may be afraid that they will share their insights and original ideas, and other individuals or groups will either capitalize on this new knowledge or take credit for it. Companies should strive to create a culture of trust and collaboration where employees feel empowered to share knowledge without fear of misappropriation. Leveraging digital technologies, such as blockchain, individuals, and businesses can create a transparent and reliable record of knowledge exchange, allowing all stakeholders to access and verify the information shared (Imeri et al., 2019). Also, employees should be actively involved in Gen-AI-powered digital transformation. Their insights and feedback should be valued since they are the ones who deal directly with processes, customers, and data (Bluemke et al., 2023). They can be the source of valuable knowledge and make suggestions regarding solving the most pressing issues with Gen-AI.

As stated above, workers may fear that their employers are looking for savings and intend to use automation to replace human workers. The fear on the part of the employee is that he or she will share their knowledge and know-how, which will be encapsulated (digitized and recorded in a permanent form) and could be multiplicated and used repetitively without the need to involve the employee who was the source of valuable knowledge. However, this perspective is not correct. Employers need talents, and they should be interested in collecting, developing, and applying their knowledge. It is important to note that knowledge sharing is not a single event, but a social process based on communication and collaboration. The knowledge quickly becomes obsolete and is useless unless applied (Jemielniak & Koźminski, 2012). Therefore, employers should change their focus from short-term profit to long-term value. Concerns about job displacement, accountability in human-machine interactions, and AI's ethical and transparent use must be addressed for employees. Managers should engage in an open dialogue with their teams about AI's role, risks, and limitations. They should also clearly communicate the role of AI systems within the company to support, rather than replace, human capacity, creativity, or decision-making.

Both firms and employees may have concerns about sharing data or information that could compromise their competitive advantage. However, new technologies, e.g. blockchain can serve as a potential solution to address trust issues. It guarantees secure, transparent data sharing, preventing tampering with data (Imeri et al. 2019) or infringing intellectual property or privacy rights. Also, as Paliszkiewicz et al. (2024) state, AI can become "a privacy protector […] to mitigate privacy risks associated with the use of Big Data by improving AI algorithms and models to better detect and prevent privacy breaches" (Paliszkiewicz et al., 2024). Incorporating AI and ML solutions can also automate the process of verifying the integrity and authenticity of shared knowledge, contributing to greater transparency. As Bluemke et al. (2023) claim, "the development of privacy-enhancing technologies has made immense progress in reducing trade-offs between privacy and performance in data exchange and analysis". Also, according to the authors, "similar tools for structured transparency could be useful for

AI governance by offering capabilities such as external scrutiny, auditing, and source verification" (Bluemke et al., 2023).

Conclusion

AI, including generative AI, optimization, and robotics can potentially increase productivity and efficiency, especially if applied to repetitive or creative tasks. However, it is important to bear in mind that the abovementioned technologies may have mixed effects on labor, particularly when emerging as part of the complex and multifaceted digital transformation process. While managers look forward to increased efficiency, productivity or cost, and time savings, employees are concerned about the impact of implementing new AI-driven solutions in their organizations and work areas.

Managers should consider the implications of introducing Gen AI on employee well-being or job security (Giuntella et al., 2023) and be aware of the strong emotions accompanying organizational changes. When introducing and promoting GenAI-based solutions, leaders should address concerns and engage employees in the transformational processes. Leaders need to empower employees, who should feel that they are subjects, not objects, of the changes taking place in organizations.

Organizations should strive to create a culture of trust and collaboration where employees feel empowered to share knowledge, insights, or feedback. Managers should be transparent about the risks involved in new ICT projects or the uncertainty of the outcomes of their innovative initiatives and transformational efforts. Cultivating a corporate culture that supports the learning process, recognizes the value of experimentation, prioritizes ethical data practices and respects employee contributions can foster and support knowledge-sharing behaviors.

Establishing trust through transparency, communication, and the right practices ensures a smooth transition into digital AI-driven transformation and successful adoption of AI technologies in the workplace. Also, if employees have a better understanding of how AI systems work and how their jobs may be affected, they are more likely to develop trust in both the new technologies and the management that implements them.

References

Ahmad, F., & Karim, M. (2019). Impacts of knowledge sharing: A review and directions for future research. *Journal of Workplace Learning*, 31(3), 207–230. https://doi.org/10.1108/JWL-07-2018-0096

Argyris, C. (1985). *Strategy, Change, and Defensive Routines*. Boston, MA: Pitman.

Ashleigh, M., Connell, C., & Klein, J. H. (2003, October). Trust and knowledge transfer: An explanatory framework for identifying relationships within communities of practice. In *The IASM Second Workshop on Trust within and between Organisations*, Amsterdam.

Barber, H. F. (1992). Developing strategic leadership: The US Army War College experience. *Journal of Management Development*, 11(6), 4–12. https://doi.org/10.1108/02621719210018208

Bennis, W. G., & Nanus, B. (1986). *Leaders: The Strategies for Taking Charge*. New York: Harper & Row.

Bluemke, E., Collins, T., Garfinkel, B., & Trask, A. (2023). *Exploring the Relevance of Data Privacy-Enhancing Technologies for AI Governance Use Cases*. ArXiv, abs/2303.08956.

Braganza, A., Chen, W., Canhoto, A., & Sap, S. (2021). Productive employment and decent work: The impact of AI adoption on psychological contracts, job engagement and employee trust. *Journal of Business Research, Elsevier*, 131(C), 485–494.

Brilman, J. (2002). *Nowoczesne koncepcje i metody zarządzania*. Warsaw: PWE.

Bulińska-Stangrecka, H. (2018). Wpływ zaufania organizacyjnego na dzielenie się wiedzą w organizacjach IT. *Studia ekonomiczne. Zeszyty Naukowe Uniwersytetu Ekonomicznego w Katowicach*, 359, 2018.

Chatterjee, S., Chaudhuri, R., & Vrontis, D. (2022). Knowledge sharing in international markets for product and process innovation: The moderating role of firm's absorptive capacity. *International Marketing Review*, 39(3), 706–733. https://doi.org/10.1108/IMR-11-2020-0261.

Coyle-Shapiro, J. A.-M., & Parzefall, M. (2008). Psychological contracts. In Cooper, C. L. & Barling, J. (eds.), *The SAGE Handbook of Organizational Behavior* (pp. 17–34). London: Sage Publications.

Davenport, T. H., & Prusak, L. (1998). *Working Knowledge: How Organizations Manage What They Know*. Boston, MA: Harvard Business School Press.

Deng, P., & Lu, H. (2022). Transnational knowledge transfer or indigenous knowledge transfer: Which channel has more benefits for China's high-tech enterprises? *European Journal of Innovation Management*, 25(2), 433–453. https://doi.org/10.1108/EJIM-07-2020-0269.

Dombrowski, C., Kim, J. Y., Desouza, K. C., Braganza, A., Papagari, S. S., Baloh, P., & Jha, S. (2007, July 1). Elements of innovative cultures. *Wiley*, 14(3), 190–202. https://doi.org/10.1002/kpm.279.

Gardner, W. L., Avolio, B. J., Luthans, F., May, D. R., & Walumbwa, F. (2005). "Can you see the real me?" A self-based model of authentic leader and follower development. *The Leadership Quarterly*, 16(3), 343–372. https://doi.org/10.1016/j.leaqua.2005.03.003.

George, A. Shaji. (2024). *Artificial Intelligence and the Future of Work: Job Shifting Not Job Loss*. vol. 2. 17–37. 10.5281/zenodo.10936490.

Gilmour, P., Coffey, B., & O'Toole, K. (2015). Trust and knowledge exchange in coastal settings. *Australian Journal of Maritime & Ocean Affairs*, 7(1), 66–74.

Giuntella, O., Koenig, J., & Stella, L. (2023). *Artificial Intelligence and Workers' Well-Being*. IZA Discussion Paper No. 16485, Available at SSRN: https://papers.ssrn.com/sol3/papers or https://doi.org/10.2139/ssrn.4589223

Gul, S., Jamal, W., Sher, A. (2020). Transfer of knowledge through the lens of organizational culture: A literature review. *Journal of Business & Tourism*, 6(1), 147–178. https://doi.org/10.34260/jbt.v6i1.186

Hasnain, S. S. (2017). Trust-A catalyst to knowledge transfer: Contemporary studies. *Advances in Social Sciences Research Journal*, 4, 29–40. https://doi.org/10.14738/assrj.41.2586

Hayen, R. L., Rutashobya, C. D., & Vetter, D. E. (2007). An investigation of the factors affecting data warehousing success. *Issues in Information Systems*, 8(2), 547–553.

Holste, J. S., & Fields, D. (2010). Trust and tacit knowledge sharing and use. *Journal of Knowledge Management*, 14(1), 128–140.

How AI Will Transform Our Workplaces? https://businessabc.net/how-ai-will-transform-our-workplaces

Imeri, A., Agoulmine, N., Feltus, C., & Khadraoui, D. (2019, January 1). *Blockchain: Analysis of the New Technological Components as Opportunity to Solve the Trust Issues in Supply Chain Management*. Springer Nature, 474–493. https://doi.org/10.1007/978-3-030-22868-2_36

Ipe, M. (2003). Knowledge sharing in organizations: A conceptual framework. *Human Resource Development Review*, 2, 337–359. doi: 10.1177/1534484303257985

Jemielniak, D., & Koźmiński, A. K. (2012). *Zarządzanie wiedzą*. Warsaw: Wolters Kluwer.

Kempster, S. J., Iszatt-White, M., & Brown, M. (2019). Authenticity in leadership: Reframing relational transparency through the lens of emotional labour. *Leadership*, 15(3), 319–338.

Kreps, S., George, J., Lushenko, P., & Rao, A. (2023). Exploring the artificial intelligence "Trust paradox": Evidence from a survey experiment in the United States. *PLoS One*, 18(7). https://journals.plos.org/plosone/article?id=10.1371/journal.pone.0288109

Królik, R. (2015). Wiedza jako potencjał strategiczny przedsiębiorstwa rodzinnego. *Studia Ekonomiczne. Zeszyty Naukowe Uniwersytetu Ekonomicznego w Katowicach*, 222, 153–165.

Kucharska, W. (2017). Relationships between trust and collaborative culture in the context of tacit knowledge sharing. *Journal of Entrepreneurship, Management and Innovation*, 13, 61–78. https://doi.org/10.7341/20171344.

Larsson, S., & Heintz, F. (2020). Transparency in artificial intelligence. *Internet Policy Review*, 9(2), 1–16. https://doi.org/10.14763/2020.2.1469

Lazić, A., & Jović, M. (2019, January 1). Strategic digital transformation of organisations. In *Advances in Economics, Business and Management Research, 5th IPMA SENET Project Management Conference (SENET 2019)* (Vol. 108, pp. 184–188). https://doi.org/10.2991/senet-19.2019.29

Lee, S. T., & Cheng, I. (2010, December 30). Characteristics and dimensions of ethical leadership in public relations. *Journal of Public Relations Research*, 23(1), 46–74. https://doi.org/10.1080/1062726x.2010.504790

Liebeskind, J. P. (1996). Knowledge, strategy, and the theory of the firm. *Strategic Management Journal*, 17, 93–107.

Lin, C., Huang, H., & Huang, T. Y. (2020, April 6). The effects of responsible leadership and knowledge sharing on job performance among knowledge workers. *Emerald Publishing Limited*, 49(9), 1879–1896. https://doi.org/10.1108/pr-12-2018-0527

Lokuge, S., & Duan, S. (2021). Towards understanding enablers of digital transformation in small and medium-sized enterprises. In *Australasian Conference on Information Systems, ACIS 2021 Proceedings*. https://doi.org/10.48550/arXiv.2111.05989

Lokuge, S., & Duan, S. (2023, January 1). Exploring the enablers of digital transformation in small and medium-sized enterprise. In *Handbook of research on business model innovation through disruption and digitalization advances in business strategy and competitive advantage*, 2022 (pp. 136–156). IGI Global. https://doi.org/10.4018/978-1-6684-4895-3.ch008

Lucas, L. (2005). The impact of trust and reputation on the transfer of the best practices. *Journal of Knowledge Management*, 9, 87–101. https://doi.org/10.1108/13673270510610350

Mohajan, H. (2016). Sharing of tacit knowledge in organizations: A review. *American Journal of Computer Science and Engineering*, 3, 6–19.

Mirbabaie, M., Brünker, F., & Möllmann Frick, N. R. J., & Stieglitz, S. (2021). The rise of artificial intelligence – understanding the AI identity threat at the workplace. *Electron Markets*, 32, 73–99. https://doi.org/10.1007/s12525-021-00496-x

Paliszkiewicz, J. (2007). Dzielenie się wiedzą oraz zaufanie w małych i średnich przedsiębiorstwach. *Zeszyty Naukowe SGGW - Ekonomika I Organizacja Gospodarki Żywnościowej*, 62, 143–152. https://doi.org/10.22630/EIOGZ.2007.62.11

Paliszkiewicz, J. (2010). Zarządzanie zaufaniem i dzielenie się wiedzą - przegląd literatury. *Prace Naukowe / Akademia Ekonomiczna w Katowicach. Technologie wiedzy w zarządzaniu publicznym*, 10, 315–322.

Paliszkiewicz, J., Chen, K., & Gołuchowski, J. (2024). *Privacy, Trust and Social Media*. New York: Routledge, Taylor & Francis.

Paliszkiewicz, J., & Gołuchowski, J. (2025). Trust in artificial intelligence – future directions. In *Trust and Artificial Intelligence*. New York: Routledge. Taylor & Francis.

Riedl, R. (2022). Is trust in artificial intelligence systems related to user personality? Review of empirical evidence and future research directions. *Electron Markets*, 32, 2021–2051.

Rolland, N., & Chauvel, D. (2000). Knowledge transfer in strategic alliances. In Despres, C. & Chauvel, D. (eds.), *Knowledge Horizons: The Present and Promise of Knowledge Management* (pp. 225–236), Boston, MA: Butterworth-Heinemann.

Sankowska, A. (2013). Relationships between organisational trust, knowledge transfer, knowledge creation, and firm's innovativeness. *The Learning Organisation*, 20(1), 85–100.

Sayyadi, M., & Provitera, M. J. (2023, March 5). *Resource-Based and Knowledge-Based Theory of Competitive Advantage: How to Compete on Resources*. https://www.europeanfinancialreview.com/resource-based-and-knowledge-based-theory-of-competitive-advantage-how-to-compete-on-resources/.

Singh, T., & Premarajan, R. K. (2007). Antecedents to knowledge transfer: Trust and culture. *South Asian Journal of Management*, 14(1), 93–104.

Soni, V. (2023). Impact of generative AI on small and medium enterprises' revenue growth: The moderating role of human, technological, and market factors. *RCBA-2023*, 6(1), 133–153.

Sreckovic, M., & Windsperger, J. (2012, January 6). The impact of trust on the choice of knowledge transfer mechanisms in clusters. In Ehrmann, Th., Windsperger, J., Cliquet, G., & Hendrikse, G. (eds.), *Network Governance: Alliances, Cooperatives and Franchise Chains* (pp. 73–85). Berlin Heidelberg: Springer Verlag, 2013. Available at SSRN: https://papers.ssrn.com/sol3/papers.cfm?abstract

Stonehouse, G., & Konina, N. Y. (2020, January 1). Management challenges in the age of digital disruption. https://doi.org/10.2991/aebmr.k.200201.001

The Six Primary Dimensions for Data Quality Assessment. Defining Data Quality Dimensions. Available at: https://www.sbctc.edu/resources/documents/colleges-staff/commissions-councils/dgc/data-quality-deminsions.pdf

Trenerry, B., Chng, S., Wang, Y., Suhaila, Z. S., Lim, S. S., Lu, H., & Oh, P. H. (2021, March 23). Preparing workplaces for digital transformation: An integrative review and framework of multi-level factors. *Frontiers Media*, 12. https://doi.org/10.3389/fpsyg.2021.620766

Usoro, A., Sharratt, M. W., Tsui, E., & Shekhar, S. (2007). An investment into trust as an antecedent to knowledge sharing in virtual communities of practice. *Knowledge Management Research and Practice*, 5, 199–212.

Vial, G. (2019, June 1). Understanding digital transformation: A review and a research agenda. *Elsevier BV*, 28(2), 118–144. https://doi.org/10.1016/j.jsis.2019.01.003

Warner, K., & Wäger, M. (2019, June 1). Building dynamic capabilities for digital transformation: An ongoing process of strategic renewal. *Elsevier BV*, 52(3), 326–349. https://doi.org/10.1016/j.lrp.2018.12.001

Wiśniewski, K. (2019). *Koncepcje zarządzania wiedzą w przedsiębiorstwie innowacyjnym*. Uniwersytet Jana Kochanowskiego, Kielce.

Xia, Z., & Yang, F. (2020, November 9). Ethical leadership and knowledge sharing: The impacts of prosocial motivation and two facets of conscientiousness. *Frontiers Media*, 11. https://doi.org/10.3389/fpsyg.2020.581236

Yeboah, A. (2023). Knowledge sharing in organization: A systematic review. *Cogent Business & Management*, 10, 2195027. https://doi.org/10.1080/23311975.2023.2195027

Zhang, Y., & Jin, S. (2023, July 11). How does digital transformation increase corporate sustainability? The moderating role of top management teams. *Multidisciplinary Digital Publishing Institute*, 11(7), 355–355. https://doi.org/10.3390/systems11070355

Zheng, W., Yang, B., & Mclean, G. N. (2010). Linking organizational culture, structure, strategy, and organizational effectiveness: Mediating role of knowledge management, *Journal of Business Research*, 63(7), 763–771.

What is a psychological contract: Types with examples. https://www.aihr.com/blog/psychological-contract/

Index

For Product Safety Concerns and Information please contact our EU representative GPSR@taylorandfrancis.com
Taylor & Francis Verlag GmbH, Kaufingerstraße 24, 80331 München, Germany

www.ingramcontent.com/pod-product-compliance
Lightning Source LLC
LaVergne TN
LVHW021126300926
849160LV00010B/885
* 9 7 8 1 0 3 2 9 5 8 7 6 7 *